T0230499

Lecture Notes in Civil Engineering

Volume 320

Lecture Notes in Civil Engineering (LNCE) publishes the latest developments in Civil Engineering—quickly, informally and in top quality. Though original research reported in proceedings and post-proceedings represents the core of LNCE, edited volumes of exceptionally high quality and interest may also be considered for publication. Volumes published in LNCE embrace all aspects and subfields of, as well as new challenges in, Civil Engineering. Topics in the series include:

- Construction and Structural Mechanics
- Building Materials
- Concrete, Steel and Timber Structures
- Geotechnical Engineering
- Earthquake Engineering
- Coastal Engineering
- Ocean and Offshore Engineering; Ships and Floating Structures
- Hydraulics, Hydrology and Water Resources Engineering
- Environmental Engineering and Sustainability
- Structural Health and Monitoring
- Surveying and Geographical Information Systems
- Indoor Environments
- Transportation and Traffic
- Risk Analysis
- Safety and Security

To submit a proposal or request further information, please contact the appropriate Springer Editor:

- Pierpaolo Riva at pierpaolo.riva@springer.com (Europe and Americas);
- Swati Meherishi at swati.meherishi@springer.com (Asia—except China, Australia, and New Zealand);
- Wayne Hu at wayne.hu@springer.com (China).

All books in the series now indexed by Scopus and EI Compendex database!

Khalid Al Marri · Farzana Mir · Solomon David ·
Aos Aljuboori
Editors

BUiD Doctoral Research Conference 2022

Multidisciplinary Studies

 Springer

Editors
Khalid Al Marri
The British University in Dubai
Dubai, United Arab Emirates

Farzana Mir
The British University in Dubai
Dubai, United Arab Emirates

Solomon David
The British University in Dubai
Dubai, United Arab Emirates

Aos Aljuboori
The British University in Dubai
Dubai, United Arab Emirates

ISSN 2366-2557   ISSN 2366-2565 (electronic)
Lecture Notes in Civil Engineering
ISBN 978-3-031-27464-0   ISBN 978-3-031-27462-6 (eBook)
https://doi.org/10.1007/978-3-031-27462-6

This Springer imprint is published by the registered company Springer Nature Switzerland AG
The registered company address is: Gewerbestrasse 11, 6330 Cham, Switzerland

Preface

In a world inundated with information and the publication of what seems like yet another conference proceedings, we struggle to find value and use for new knowledge coming from young researchers and emerging minds. The question we then need to ask ourselves is how this publication engage and reinvigorate its users and renews the relationship between scholarship and its application in a world that needs that dynamic interaction between praxis and theory. This proceedings being multidisciplinary in nature is the compilation of the efforts of our student researchers looking at a variety of issues from various perspectives. The diversity in the presentation of ideas is representative of its inert value in the creation of innovative solutions for varied areas of application and industries.

Since 2015, the British University in Dubai has been hosting the BUiD Doctoral Research Conference (BDRC), making it the pioneer in organising such a large doctoral research conference in the UAE. Many internationally renowned experts and academicians in the fields of Computer Science, Education and Engineering have participated in these conferences and have made innovative and outstanding presentations. The visitors and participants of the conference include experts, academics, engineers and other professionals from well-known and leading institutions of higher learning, research and enterprise.

The 6th BUiD Doctoral Research Conference 2022 focused on providing students with a research forum through which their work can be shared and published. Inherently multidisciplinary in nature, it is an opportunity to discuss research and training in education, business, law, engineering, sustainable design and innovation. The conference is intended for upcoming researchers and those who carry the dual role of researcher and educator, thought leader, practitioner, technocrat or business leader. We hope to bring together a diverse group to present and discuss the latest trends, discoveries, praxis and interests in these multidisciplinary areas.

Some of the papers delved into pertinent issues related to education, particularly the reformation of the education system and the movement away from traditional type approaches to approaches that contribute to sustainable paths and development in line with future goals. In a world which has become highly dependent on social media as a means of global communication, other papers explored the data aspect of social media platforms and the ability to organise and make sense of it in a global multicultural environment. In the area of ICT and innovation, valuable connections were made to the Fourth Industrial Revolution, the requirements that would ensure its success and the impact it would have on Middle Eastern development. Issues of material and structural performance were also a theme explored under the overarching theme of climate in the Middle East and its impact on the materials used in structural development and the contributing factors and effects on international best practices in the construction industry.

The British University in Dubai is extremely proud of the contribution it makes to the development of knowledge specifically as it pertains to Middle Eastern issues and the contribution the region makes to solving wider global issues in various spheres and disciplines of academia. This conference proceedings is committed to providing young researchers with the forum to develop new ideas and the promotion of a rigorous research environment in the areas of education, innovation, business, sustainability, computer science and engineering. All papers have been rigorously reviewed and have been presented in multiple iterations before being accepted for this inaugural issue. According to their subject and their merit by professionals, eighteen of the best papers presented at the BDRC 2022 Conference are included in this issue.

We thank all of the authors who contributed to this issue.

Contents

The Knowledge-Evolving Project: Fundamentals to a Complete Project Knowledge Management Discipline 1
Ghassan Dabbour

The Impact of Macroeconomic Factors on the Nifty Auto Index 11
Rhoda Alexander and Husam Aldin Al-Malkawi

Ensuring the Fourth Industrial Revolution's Connectivity Requirements in the UAE 23
Khalid Alawadhi and Khalid Almarri

Enhancing Products Delivery Through the Application of Innovative Operating Model Based on Hybrid Agile Delivery Method: Case Information Communication Technologies "ICT" Service Providers 35
Madhad Ali Said Al Jabri

Successful Practices of Leadership on School Improvement: A Case Study in a Private School in Dubai 47
Ahmed ElSayed Abouelanein and Mohamed Hossni

Educating Learners with SEND in One Private Mainstream School in Dubai: Effectiveness and Challenges of Policy Implementation 57
Muntaha Badawieh, Abdulai Abukari, and Eman Gaad

The Role of Lexical Cohesion in Improving Twelfth Graders' Essay Writing Quality 67
Abdelhamid A. Khalil, Emad A. S. Abu-Ayyash, and Sa'Ed M. I. Salhieh

Critical Thinking Skills Profile of High School Students in AP Chemistry Learning 79
Gilan Raslan

Moving from the Subject-Based Curriculum to the Skills-Based
Curriculum in Abu Dhabi Schools: Does It Lead to Reform?
A Theoretical Analysis & Case Study Paper in One of Abu Dhabi
Private Schools 97
Yaser Abdulrahman Ibrahim

The Education System in Post-conflict Syria – Examining PIRLS
as an International Assessment Measure to Ensure the Quality
of Students Achievements 107
Yaser Abdulrahman Ibrahim

Using Nearpod to Promote Engagement in Online ESL Classes:
A Mixed-Methods Study in the Context of Higher Education 117
Azza Alawadhi and Rawy A. Thabet

Investigating Emirati Students' Practices of Learning Block-Based
Programming in an Online Learning Context: Evidence
from a UAE National Program 131
Wafaa Elsawah and Tendai Charles

A Quantitative Study on the Impact of Online Learning on Reading
Comprehension Skills 143
Ranya Ahmed El Haddad and Sa'Ed Mohammad Issa Salhieh

Simulation Study on the Effect of Courtyards Design on Natural
Ventilation: The Case Study of a Beauty Centre in Germany 155
Bana Eid and Hanan M. Taleb

A Sustainable Approach to Improve the Interior Design of Existing
Space: The Case of the BUiD Main Lobby 167
Rahaf Aloudeh, Manar Elmardi, and Wael Sheta

Summarising a Twitter Feed Using Weighted Frequency 179
Zina Ahmed Abohaia and Yousef Mamdouh Hassan

Preventive Maintenance Using Recycled Asphalt: Review 189
Aishah H. O. Al Shehhi, Gul Ahmed Jokhio, and Abid Abu-Tair

Overview of Concrete Deterioration Due to Sulphate Attack
and Comparison of Its Requirements in International Codes 199
Diala Basim Al-Haddad, Gul Ahmed Jokhio, and Abid Abu Tair

The Knowledge-Evolving Project: Fundamentals to a Complete Project Knowledge Management Discipline

Ghassan Dabbour

Abstract This paper establishes a knowledge management doctrine unique to the field of project management. The paper investigates and expands on the theoretical frameworks and practices of organisational knowledge management – which is the dominate doctrine in the discipline of knowledge management – to deduce a theory and a practice of project knowledge management that is distinctive from other knowledge management fields. It is discovered that the knowledge dynamics at the foundational level of project knowledge management is a symmetrical transposition of that of organisational knowledge management. Thus, the *Knowledge-Evolving Project* doctrine is established as the inverse of the *Knowledge-Creating Company*. Exploring a knowledge management discipline distinctive to project management inevitably leads to new insights that are potentially significant to managing knowledge in projects in particular and to knowledge management discipline in general. Most of the current literature on project knowledge management is almost indistinguishable from organisational knowledge management. The very same theoretical and empirical concepts and ideas are applied interchangeably in both management fields. This paper takes the rare step of creating a new theory and practice of project knowledge management such that it is distinctive from other knowledge management fields.

Keywords Project knowledge management · Knowledge management · Knowledge exploration · Knowledge exploitation

1 Introduction

A complete and inclusive project knowledge management archetype has been attempted several times before; some of the most notable examples include Gasik (2011), Koskinen and Pihlanto (2008), and Kasvi et al. (2003). Such studies, however, readily take the well-acquainted concepts of organisational knowledge management

G. Dabbour (✉)
The British University in Dubai, Dubai, United Arab Emirates
e-mail: dabbour.ghassan@gmail.com

© The Author(s) 2023
K. Al Marri et al. (eds.), *BUiD Doctoral Research Conference 2022*,
Lecture Notes in Civil Engineering 320,
https://doi.org/10.1007/978-3-031-27462-6_1

and absorb them into a project management framework. There are thus hardly any foundational differences between managing knowledge in projects and managing knowledge in organisations.

There are several reasons why project knowledge management is entangled with organisational knowledge management. Perhaps, though, the most evident of them is that at the fundamental level of knowledge management, there is a highly influential organisational knowledge management paper that arguably popularised the whole field. Written by Ikajiro Nonaka and published by the Harvard Business Review in 1991, the *Knowledge-Creating Company* lays down the basics of knowledge management in organisations. These basics, from 'knowledge dimensions', to 'knowledge management processes', and to 'knowledge transformations', indiscriminately infiltrate the project knowledge management field. Exploring a knowledge management discipline idiosyncratic to project management can be potentially theoretically and empirically rewarding to managing knowledge in projects in particular and to the field of knowledge management in general.

The proposed 'Knowledge-Evolving Project' is thus meant to replace the *Knowledge-Creating Company* as the cornerstone upon which a complete knowledge management discipline distinctive to project management can be built. Simply put, the Knowledge-Evolving Project is an inverse of the knowledge-creating company. While the knowledge-creating company encourages a *typical* organisation to utilise knowledge and knowing to *explore* beyond constraints to engulf more markets and industries, the inverse of that, the Knowledge-Evolving Project, encourages a *typical* project to utilise knowledge and knowing to effectively and efficiently *exploit* predefined constraints to meet on budget, time, and scope. The word 'typical' here is italicised for a reason; it is meant to describe a classical modernist definition of projects and organisations, during which time each were clearly distinctive.

The objectives of this research paper are straightforward: (1) to untangle project knowledge management from organisational knowledge management, (2) to introduce a new knowledge management doctrine idiosyncratic to project management, and (3) to provide guidelines and pathways to further develop this project knowledge management doctrine. Accordingly, this paper is divided into three main sections: the first section will undo the theoretical entanglement between project knowledge management and organisational knowledge management and by extend pave the way for the doctrine of the Knowledge-Evolving Project; the second section will strengthen the Knowledge-Evolving Project doctrine by unpacking the resulting practical model; and the final section will present ideas and pathways for further research.

2 Unknotting Project Knowledge Management from Organisational Knowledge Management

2.1 A Project or an Organisation?

Project management is often celebrated as a modern era institution, but the fact remains that the ancients weren't exactly unaware of its methods and techniques. Some of the amazing wonders of ancient civilisations such as the Gardens of Babylon, the pyramids of Giza, and the Colossus of Rhodes were project-based construction activities. Project management as we know it today could be traced back to Henry Gantt and Frederic Taylor who introduced the elements of science into management techniques during the early 1900s. Besides scientific models such as PERT analysis and the Critical Path Method, the classical modernist era of project management determined its identity by the iron triangle of budget, scope, and time. Perhaps the most significant event that solidified the identity of the project management discipline is the formation of project management professional associations such as the PMI Project Management Institute and APM Association of Project Management.

During the postmodern era of the 1960–1990s, the discipline of project management was permeable to the newly emerging social approach of management in business studies and economics such as organisational behaviour, strategic management, and game theory. As a result, project management was becoming more complex in theory and practice as it pushed its identity borders. It was during this era that the concept of 'projectification' and the 'project-based organisation' became common knowledge. Project management continued to push its boundaries, covering operations that used to be beyond its reach – horizontally stretching from pre-planning to post-completion, and vertically zooming outwards from project programs to project portfolios (Morris 2013). After the 1990s, during the early post postmodern era, a new school of thought began redefining the very identity of project management by challenging the iron triangle constraints. The definition of project management was gradually changing from 'a temporary endeavor' to 'a temporary organisation' (Winter et al. 2006). This approach, popularly known as the 'Scandinavian Turn', is being championed by many high-profile thinkers of project management such as Hodgson and Cicmil (2016), Jacobsson et al. (2016), and Svejvig and Grex (2016), (see Table 1).

While this expansion is exciting and promising, project management as a discipline is losing its clearly bounded identity. It is becoming increasingly difficult to differentiate between a project and an organisation. The project in the organisation perspective (Anderson 2010), the project-based organisation (Prado and Sapsed 2016), and the projectification of organisations (Packendorff and Lindgren 2014) are blending organisations and projects in theory and practice. Thus, investigating a knowledge management subject matter exclusively idiosyncratic to project management is already a doomed mission should the investigation adopt the post postmodern or the postmodern definition of projects and project management. The first step to

Table 1 Project management through the ages

Pre-Modern Era Historic	• Projects of the ancients were primarily construction centred such as Great Wall of China, Colossus of Rhodes, and Gardens of Babylon
Modern Era 1940–1970s	• Scientific and engineering models (e.g., PERT analysis and Gantt Charts) • Scope, time, budget constraints and life cycle • Project management associations formed
Postmodern Era 1960–1990s	• Organisational behaviour, game theory, knowledge management, and other social and economic centred sciences penetrate the project management field • Projectification • Project-based organisations
Post Postmodern Era 1990s – Today	• Project temporality and constraints challenged • Scandinavian turn: projects as temporary organisations Rethinking Project Management, Scandinavian School of Project Management, Making Projects Critical

approach this mission is therefore to adopt the classical modernist (typical) perspective that clearly and precisely differentiates projects from organisations. Thus, a project here is defined as the temporary endeavour with a predefined constraint of time, scope, and budget (PMI 2013). Projects are typically unstable, terminal, and operate within a fluctuating environment and wrestling with constraints. An organisation on the other hand is an institution that is typically lasting, mostly stable, and works within a routine environment.

2.2 Knowledge Dynamics in Projects and Organisations

The second step is to identify where the characteristics of knowledge activities in projects and knowledge activities in organisations diverge. It is at this point of divergence that a knowledge management discipline unique to either fields can be identified. Fortunately, because organisational knowledge management already has a doctrine – Nonaka and Takeuchi's (1995) *The Knowledge-Creating Company* – this task is made substantially easier. We know from Nonaka & colleagues' writings (Nonaka and Von Krogh 2009; Nonaka and Tomaya 2005; Nonaka and Nishiguchi 2001) that individuals in a typical organisation are encouraged to come up with innovations that destabilises the monotonous environment and induces change. It is only through this novelty of knowledge creation that organisations can push its constraints outwards to engulf more industries and markets and become more prosperous and powerful. In organisational learning, there is a term for such a knowledge dynamic; *knowledge exploration.* Knowledge exploration is the dexterity of rethinking away from established knowledge in previously unanticipated ways (Hatch and Cunliffe 2013).

On the other hand, individuals in a typical project are encouraged to specialise in and build on the full knowledge predefined at the project outset to calm the fluctuating environment and induce stability. It is only through this novelty of *evolving* the knowledge established at initiation to encounter unforeseen opportunities and risks that projects can meet their predefined constraints efficiently and effectively. Again, in organisational learning, there is a term for such a knowledge dynamic; *knowledge exploitation*. Knowledge exploitation is the dexterity of using the existing knowledge base to fully develop and evolve said knowledge for the application it was meant to address (Hatch and Cunliffe 2013).

So it would seem that while innovative exploration overtakes innovative exploitation in organisations, innovative exploitation overtakes innovative exploration in project settings. The correlation between the knowledge dynamics of an organisation and a project is thus an inverse function; if one is represented as a spiral that circles outwards to explore more markets and more industries, the other should be represented as a spiral that circles inwards to exploit the predefined knowledge and meet the budget, time, and scope constraints. The knowledge dynamics of each formulate the basic theoretical background upon which knowledge management should be built. *The Knowledge-Creating Company* does just that – it adopts specific knowledge practices and processes that are associated with knowledge exploration. Likewise, the proposed Knowledge-Evolving Project should adopt specific knowledge practices and processes that are associated with knowledge exploitation.

2.3 The Epistemology of Tacit Knowing

The third and last step to untangling project knowledge management from organisational knowledge management is subscribing project knowledge management to a different epistemological perspective than the one assumed by organisational knowledge management. This step is important because the current literature on organisational knowledge management is struggling with the philosophical complexities, ambiguities, and shortcomings of the concept of 'tacit knowing'.

First introduced by Hungarian-British chemist turned epistemologist – Michael Polanyi –, tacit knowing refers to a type of knowledge best explained in his own words: 'we know more than we can tell'. Nonaka and Takeuchi (1995) explain that where tacit knowledge is the kind of knowledge that is near impossible to articulate, 'explicit' knowledge is the kind of knowledge that is easily expressible in language. They then continue to demonstrate how subjective tacit knowledge can be externalised into objective explicit, and then internalised back into subjective tacit. The problem here is that all explicit knowledge is *tacitly rooted* (Polanyi 1969). After all, how can knowledge exist independently of the conscious knower(s)? This makes the concept of 'knowledge transformation' from one form to another very problematic (Crane and Bontis 2014; Garrick and Chan 2017; Ribeiro, 2013). Nonaka and colleagues' replies to criticism (Nonaka and Nishihara 2018; Nonaka and Von Krogh

2009; Nonaka et al. 2014) introduced even more complexities and philosophical ambiguities.

This problem could be solved by substituting 'transformation' with 'utilisation'. This means that what seems to us like transformation from one form of knowledge to another is actually one form of knowledge being utilised to enhance the epistemological action of the other. So, for example, one can use the explicit knowledge publicly available about the mechanics of engines to enhance the creative application of their tacit knowledge to accelerate the production line of car engines. No transformation occurs here, only utilisation and enhancement. This idea dissolves all the philosophical complexity associated with tacit knowing by moving away from knowledge ontology to knowledge application.

2.4 The Knowledge-Evolving Project Emerges

The discussion of the previous three subsections has produced a new construct of project knowledge management that is idiosyncratic to project management. Projects are temporary endeavours limited in time, scope, and budget. In project environments, knowledge exploitation overtakes knowledge exploration to induce stability to fluctuations. Thus project knowledge management is the utilisation of explicit or tacit forms of knowledge to enhance the epistemological action of other explicit and tacit forms of knowledge in project environments to meet projects' constraints. This is the Knowledge-Evolving Project (Table 2). The next section will unpack what this really means in practice.

Table 2 The knowledge-evolving project and the knowledge-creating company

	The knowledge-evolving project	The knowledge-creating company
Knowledge dynamics	Knowledge exploitation overtakes knowledge exploration	Knowledge exploration overtakes knowledge exploitation
Knowledge function	Stabilisation of fluctuating environment	Destabilisation of routine environment
Knowledge processes	Evolution of predefined knowledge	Exploration of new knowledge
Knowledge management objective	Successfully meeting constraints	Growing beyond constraints
Epistemological perspective	Knowledge utilisation	Knowledge transformation

3 The FRDA Model of the Knowledge-Evolving Project

Nonaka and Takeuchi's (1995) knowledge creation model of the Knowledge-Creating Company is called the SECI model. SECI is an acronym for the knowledge processes of Socialisation, Externalisation, Combination, and Internalisation. Socialisation describes epistemological practices such as experience, experimentation, and observation to transform the tacit knowledge of individuals to the tacit knowledge of other individuals; Externalisation describes epistemological practices such as discussion and dialogue to transform the tacit knowledge of individuals to explicit knowledge of groups; Combination describes epistemological practices such as documentation and information technology to transform the explicit knowledge of groups to explicit knowledge of other groups; and finally, Internalisation describes epistemological practices such as training and mentoring to transform explicit knowledge of groups to tacit knowledge of individuals. For an organisation to be successful, practitioners need to move in this order and then back to Socialisation. It is important to move in this order so that effectually the epistemological action here is *knowledge exploration.* This order of movement is represented by a spiralling circle starting with Socialisation of individuals and forever expanding outwards for the lasting organisation to dominate more markets and industries.

A proposed knowledge evolution model of the Knowledge-Evolving Project should effectually induce the epistemological action of *knowledge exploitation* of predefined complete knowledge. Since knowledge exploitation is an inverse of knowledge exploration, all must do is invert the SECI model. The proposed FRDA model is just that – an inversion of the SECI model: (1) where SECI's spiral starts with tacit knowledge of individuals, FRDA's spiral starts with explicit knowledge of stakeholder groups; (2) where SECI's spiral goes forever for a lasting organisation, FRDA's spiral terminates at the centre point where the project is complete; (3) where SECI spirals outwards for knowledge creation, FRDA spirals inwards for knowledge evolution; and (4) where SECI's spiral starts with incomplete partial tacit knowledge, FRDA's spiral starts with a complete set of explicit predefined knowledge.

FRDA is an acronym for the knowledge processes of Formalisation, Realisation, Deconstruction, and Assimilation. Formalisation describes epistemological practices such as identification, codification, authorisation, and information technology that utilises the explicit knowledge of groups to enhance the explicit knowledge of other groups; Realisation describes epistemological practices such as implementation and application of predefined knowledge into reality that utilises the use of the explicit knowledge of groups to enhance the tacit knowledge of individuals working on their particular parts of the project; Deconstruction describes epistemological practices such as creativity, innovation, sense-making, and breakdown of predefined knowledge so that it is malleable enough to confront the unforeseen risks and opportunities of reality that utilises the use of tacit knowledge of individuals to enhance the tacit knowledge of other individuals; and finally, Assimilation describes epistemological

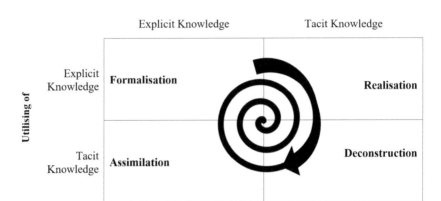

Fig. 1 The FRDA model

practices such as knowledge updating, modernisation, and restructuring of the predefined knowledge that utilises the use of tacit knowledge of individuals to enhance
the explicit knowledge of groups. For a project to be successful, practitioners need
to move in this order and then back to Formalisation. It is important to move in this
order so that effectually the epistemological action here is knowledge exploitation
(Fig. 1).

4 Conclusion and Further Research

This study has only scratched the surface of the Knowledge-Evolving Project and
already new directions and ideas potentially significant for management knowledge
in projects and knowledge management in general have emerged. The idea that
knowledge utilisation substitutes knowledge transformation for example, may be
deployed to revise mature knowledge management topics such as boundary objects
and the agency of artificial intelligence in new and exciting ways.

For the post postmodern and postmodern thinkers of project management who
identify projects as temporary organisations by melting the three classical identifiers
of projects – budget, scope and time – with its surrounding organisation or environment, the FRDA model can be used to investigate their theories in new ways
by superimposing it with the SECI model. The method and manner with which the
superimposition is applied would depend on the subject matter being investigated;
FRDA and SECI could be superimposed in series, in parallel, or as a decision-making
tree.

The FRDA model and its knowledge processes were applied to projects in an
ideal and generalised sense. It is hence open for reinterpretation and remodelling to
the various dimensions and classes of different project types, such as construction,

manufacturing, business re-modelling, and research and development. Similarly, the FRDA model can be repurposed for investigating projects that are defined by their increasing level of agility and adaptability rather than a shift in institutional configuration. Clearly there is much yet to explore and study to create a complete, bounded, and well-defined project knowledge management discipline. The Knowledge-Evolving Project doctrine is merely the cornerstone.

References

Andersen, E. S. (2010). Rethinking project management–an organisational perspective. *Strat. Direct., 26*(3), 345.

Crane, L., & Bontis, N. (2014). Trouble with tacit: developing a new perspective and approach. *J. Knowl. Manag., 18*(6), 1127–1140.

Garrick, J., & Chan, A. (2017). Knowledge management and professional experience: the uneasy dynamics between tacit knowledge and performativity in organizations. *J. Knowl. Manag., 21*(4), 872–884.

Gasik, S. (2011). A model of project knowledge management. *Project Manag. J., 42*(3), 23–44.

Hatch, M. J., & Cunliffe, A. L. (2013). *Organization theory: modern, symbolic and postmodern perspectives* (3rd ed.). Oxford: Oxford University Press.

Hodgson, D., & Cicmil, S. (2016). Making projects critical 15 years on: a retrospective reflection (2001–2016). *Int. J. Manag. Proj. Bus., 9*(4), 744–751.

Jacobsson, M., Lundin, R. A., & Söderholm, A. (2016). Towards a multi-perspective research program on projects and temporary organizations: analyzing the scandinavian turn and the Rethinking effort. *Int. J. Manag. Proj. Bus., 9*(4), 752–766.

Kasvi, J. J., Vartiainen, M., & Hailikari, M. (2003). Managing knowledge and knowledge competences in projects and project organisations. *Int. J. Proj. Manag., 21*(8), 571–582.

Koskinen, U., & Pihlanto, P. (2008). *Knowledge management in project-based companies: an organic perspective.* London: Palgrave Macmillan.

Morris, P. W. (2013). *Reconstructing project management.* Hoboken: John Wiley & Sons.

Nonaka, I., & Nishiguchi, T. (2001). *Knowledge emergence: social, technical, and evolutionary dimensions of knowledge creation.* Oxford: Oxford University Press.

Nonaka, I., & Takeuchi, H. (1995). *The knowledge-creating company: how japanese companies create the dynamics of innovation.* Oxford: Oxford University Press.

Nonaka, I., & Von Krogh, G. (2009). Perspective—tacit knowledge and knowledge conversion: controversy and advancement in organizational knowledge creation theory. *Organ. Sci., 20*(3), 635–652.

Nonaka, I., & Nishihara, A.H. (2018) Introduction to the concepts and frameworks of knowledge-creating theory. In: Nishihara, A.H., Matsunaga, M., Nonaka, I., & Yokomichi, K. (eds) *Knowledge Creation in Community Development.* Basingstoke : Palgrave Macmillan

Nonaka, I. (1991). The knowledge-creating company. *Harvard Bus. Rev., 69*, 96–104.

Nonaka, I., Kodama, M., Hirose, A., & Kohlbacher, F. (2014). Dynamic fractal organizations for promoting knowledge-based transformation–a new paradigm for organizational theory. *Eur. Manag. J., 32*(1), 137–146.

Packendorff, J., & Lindgren, M. (2014). Projectification and its consequences: narrow and broad conceptualisations. *South Afr. J. Econ. Manag. Sci., 17*(1), 7–21.

PMI. A Guide to the Project Management Body of Knowledge. 5th edn. Project Management Institute (2013)

Polanyi, M. (1969). *Knowing and being: essays.* Chicago: Chicago University Press.

Prado, P., & Sapsed, J. (2016). The anthropophagic organization: how innovations transcend the temporary in a project-based organization. *Organ. Stud., 37*(12), 1793–1818.

Ribeiro, R. (2013). Tacit knowledge management. *Phenomenol. Cogn. Sci., 12*(2), 337–366.

Svejvig, P., & Grex, S. (2016). The Danish agenda for rethinking project management. *Int. J. Manag. Proj. Bus., 9*(4), 822–844.

Winter, M., Smith, C., Cooke-Davies, T., & Cicmil, S. (2006). The importance of 'process' in rethinking project management: the story of a UK government-funded research network. *Int. J. Proj. Manag., 24*(8), 650–662.

The Impact of Macroeconomic Factors on the Nifty Auto Index

Rhoda Alexander and Husam Aldin Al-Malkawi

Abstract The aim of the paper is to investigate the association between selected macroeconomic variables like crude price, exchange rate, index of industrial production, inflation, interest rate, repo rate, gold price and the auto index of the National Stock Exchange (NSE) of India during a time when the automotive sector in India witnessed the sharpest dip in sales. The study adopts Autoregressive Distributed Lag (ARDL) co-integration approach and performs suitable diagnostic tests. Results indicate that, exchange rate has a significant negative relationship with Nifty auto index in the long run. Additionally, crude price, index of industrial production and repo rates are statistically significant determinants of Nifty auto index. On the contrary, first lag of crude price is found to be a possible predictor of the index in the short run. The study provides important implications for researchers, corporations, portfolio managers, investors, and government. Despite the availability of a large body of literature exploring the association between macro-economic factors and stock market in India, research exploring the association between the former and Indian auto indices has been sparse. Hence, this study is intended to fill this gap in the literature.

Keywords Macroeconomic factors · NSE · Nifty Auto Index · ARDL co-integration · India

1 Introduction

The Indian automotive industry recorded the biggest fall in its vehicle sales (almost 40%) in August 2019 (Hindustan Times 2019). During the same time, Nifty Auto index, which tracks the performance of the automotive sector in the National Stock

R. Alexander (✉)
Department of Management, Jaipuria Institute of Management, Noida, India
e-mail: rhoda.alexander.fpm20n@jaipuria.ac.in

H. A. Al-Malkawi
Faculty of Business and Law, The British University in Dubai, Dubai, UAE
e-mail: husam.almalkawi@buid.ac.ae

© The Author(s) 2023
K. Al Marri et al. (eds.), *BUiD Doctoral Research Conference 2022*,
Lecture Notes in Civil Engineering 320,
https://doi.org/10.1007/978-3-031-27462-6_2

Exchange also fell by about the same rate in comparison to its previous year's value. Generally, a vibrant automotive sector is considered as an important indicator of the economic performance of any country (Tambade et al. 2019). Past literature states that, macroeconomic variables measure the economic stability of a country and cannot be controlled by corporations (Mohi-u-Din and Mubasher 2013); but, might affect the volatility of their stock prices (Sheikh et al. 2020). Hence, by drawing on asset pricing theory, this paper seeks to examine the association between selected macroeconomic variables and the Nifty Auto Index during a time when the automotive industry in India recorded the sharpest fall in their vehicle sales post year 2000.

Many researchers have attempted to understand the relationship between the macro economy and the stock prices in general (Mishra et al. 2010; Tripathi and Seth 2014). Except for a recent study (Alexander and Al-Malkawi 2022), there is only limited discussion in literature regarding the impact of the macroeconomic factors on the movements of the auto indices in an Indian context. Therefore, we believe that this research study will make a significant contribution to the body of knowledge. Hence, this study seeks to investigate the following question:

RQ: How do the selected macroeconomic variables like crude oil price, gross domestic product (GDP), inflation, exchange rate, interest rate and gold price affect the Nifty Auto Index?

Our study identifies the specific macroeconomic variables (like crude price, exchange rate, GDP, inflation, gold price and interest rate) affecting the auto index of NSE and the extent of the impact in terms of long run and short run relationship using Autoregressive Distributed Lag (ARDL) co-integration technique. The findings from this study will have implications for researchers, corporations, portfolio managers, investors, and policy makers.

This paper is organized as follows. A brief review of literature is provided in Sect. 2, data and methodology is explained in Sect. 3, results are illustrated in Sect. 4, and finally, conclusion along with implications and limitations are discussed in Sect. 5.

2 Literature Review

The extant literature has identified several macroeconomic indicators that affect the automobile industry around the world. For example, Shahabuddin (2009) examined the impact of various economic factors like discount rate, GDP, GNP and other leading economic indicators on automobile sales in the US using regression analysis. Both the lagged and unlagged independent variables showed same strength of relationship with the selected response variables. In another research, Srivastava (2010) investigated the impact of change in interest rate, wholesale price index and industrial production on the Indian stock market using the Johansen's co-integration test. This study revealed that Indian stock markets are influenced by domestic macroeconomic factors when compared to global factors in the long run. Gaspareniene and Remeikiene (2014) employed correlation and multiple regression analysis to determine the link between macroeconomic factors that influenced the EU automotive industry during the period of global financial crisis. The results indicated that automobile production is strongly influenced by its demand (i.e. new vehicle registration) and the GDP. It also found a moderate correlation between public debt and automobile production.

More recently, Büyük et al. (2018) studied the relationship between macroeconomic dynamics and the automobile sales among four topmost auto production countries viz China, USA, Japan, and Germany by using Ordinary Least Squares (OLS) and Fixed Effect Model (FEM). The findings revealed that real GDP, gasoline price, car production have positive linkage with car sales while change in exchange rate, GDP per capita and inflation and cause the opposite. In another research, Nanda & Panda (2018) examined the impact of firm-specific and macroeconomic indicators on the profitability of Indian manufacturing firms. The study claims that firm-specific variables and exchange rate can be considered as potential indicators of manufacturing firm profitability. However, exchange rate is no better predictor in the short run when compared to the long run. Furthermore, Misra (2018) sought to examine the possibility and strength of linkages between Sensex and a few macroeconomic factors using various statistical tests like co-integration, granger causality and vector error correction methods. They found that there is a long-run causal relationship between money supply, inflation, index of industrial production (IIP), gold prices, interest rates, exchange rate, foreign institutional investment, and BSE Sensex. Additionally, the study also identified a short run causality between inflation and Sensex.

As discussed above, there had been many studies done in the past connecting stock market performance and macroeconomic indicators in many countries including India. However, there exist a dearth of research specifically aimed at finding a possible nexus between these macroeconomic factors and the auto indices in the Indian stock market. Moreover, the above studies have reported inconsistent results. Thus, this research contributes to the existing literature by analyzing the possible long run and short run relationship between selected macro-economic variables and the Nifty Auto index by employing an ARDL cointegration method.

3 Data and Methodology

3.1 Data

The review of literature was followed by a discussion with an industry expert to identify crude oil price, exchange rate, gross domestic product, inflation, gold price and repo rate as the independent variables of the study. All variables except repo rate were converted to their logarithmic values. A brief discussion on the data and their sources are provided in Table 1. This covers the period January 2017 to August 2019.

Table 1 Description of data

	Variables	Description	Source (Website)
1	NiftyAuto	NSE auto index	National Stock Exchange
2	CRUDE	Spot price of Brent crude (dollars per barrel)	US Energy Information Administration
3	ER	Exchange rate (USD-INR)	Reserve Bank of India (RBI)
4	IIP	Index of Industrial Production is used as a proxy for GDP	RBI
5	REPO	Interest rate	RBI
6	CPI	Consumer Price Index for inflation	RBI
7	GoldPrice	Gold prices ($/troy oz)	World Bank

3.2 Methodology

Auto Regressive Distributed Lag (ARDL) co-integration method (Pesaran et al. 2001) is employed to empirically analyze the long run linkage and dynamic interaction between the macro variables and the Nifty auto index.

3.3 Ardl Model Specification

The ARDL models constructed to empirically analyze the impact of macroeconomic variables on the movement of Nifty Auto index is given below (Joshi and Giri 2015; Alexander and Al-Malkawi 2022).

$$
\begin{aligned}
\Delta \text{LNiftyAuto} &= A_0 + A_1 \text{LCRUDE}_{t-1} + A_2 \text{LER}_{t-1} + A_3 \text{LIIP}_{t-1} \\
&+ A_4 \text{REPO}_{t-1} + \sum_{i=1}^{q} a_i \Delta \text{LNifty Auto}_{t-i} + \sum_{i=1}^{q} b_i \Delta \text{LCRUDE}_{t-i} \\
&+ \sum_{i=1}^{q} c_i \Delta \text{LER}_{t-i} + \sum_{i=1}^{q} d_i \Delta \text{LIIP}_{t-i} + \sum_{i=1}^{q} e_i \Delta \text{REPO}_{t-i} + \varepsilon_t
\end{aligned} \tag{1}
$$

Here, the first part of this equation with A_1, A_2, A_3 and A_4 refer to the long run coefficients and the second part with a_i, b_i, c_i, d_i and e_i refers to the short run coefficients.

The null hypothesis is stated as, H_0: $A1 = A2 = A3 = A4 = 0$ (i. e no co-integration) and the alternate hypothesis is specified as, $H_1 = A1 \neq A2 \neq A3 \neq A4 \neq 0$ (i.e. co-integration).

The prefix L indicates that the model uses data in the log form. As there was an indication of significant multicollinearity between REPO and LGoldPrice ($\varrho = -0.84$) from the pre-generated correlation matrix, the latter was eliminated from subsequent analyses.

3.4 ARDL Bounds Testing

In the ARDL procedure, the first step is to estimate the above Eq. (1) by an Original Least Square regression to test the possibility of a long run relationship between the corresponding variables. This is done by conducting an F-test which tests for the joint significance of the lagged levels of variables. After establishing co-integration, the next step is to estimate the conditional ARDL long run model.

Similarly, the conditional ARDL model for LNiftyAuto is specified as follows:

$$\Delta \text{LNiftyAuto}_t = A_0 + \sum_{i=1}^{q} A_1 \Delta \text{LNifty Auto}_{t-i} + \sum_{i=1}^{q} A_2 \Delta \text{LCRUDE}_{t-i} + \sum_{i=1}^{q} A_3 \text{LER}_{t-i} + \\ + \sum_{i=1}^{q} A_4 \text{LIIP}_{t-i} + \sum_{i=1}^{q} A_5 \Delta \text{REPO}_{t-i} + \varepsilon_t$$

$$(2)$$

In the final step, the short run dynamic parameters are estimated by an error correction model with the long run estimates. The error correction version of the above model is as follows:

$$\Delta \text{LNiftyAuto}_t = \text{const} + \sum_{i=1}^{q} a_i \Delta \text{LNifty Auto}_{t-i} + \sum_{i=1}^{q} b_i \Delta \text{LCRUDE}_{t-i} + \sum_{i=1}^{q} c_i \Delta \text{LER}_{t-i} + \\ + \sum_{i=1}^{q} d_i \Delta \text{LIIP}_{t-i} + \sum_{i=1}^{q} e_i \Delta \text{REPO}_{t-i} + c_1 \text{ECM}_{t-1} + \varepsilon_t$$

$$(3)$$

a_i, b_i, c_i, d_i and e_i are the short run dynamic coefficients to equilibrium and c_1 is the speed adjustment coefficients of LNiftyAuto.

4 Results and Discussion

4.1 Descriptive Statistics

A descriptive summary of all the variables is provided in the Table 2 below. No major discrepancies are observed.

4.2 Augmented Dickey Fuller (ADF) Test

Before examining the association between selected macroeconomic factors and the Nifty Auto index, their stationarity properties were assessed using unit root tests like ADF tests and Philips Perron (PP) tests. The results of ADF tests are provided in Table 3 below.

Table 2 Descriptive statistics of all variables

Variables	Mean	Std. Dev	Variance	Skewness	Kurtosis	Min	Max
LNiftyAuto	9.203	0.141	0.020	−0.754	2.769	8.826	9.395
LCrude	4.136	0.148	0.022	−0.159	2.035	3.848	4.388
LER	4.212	0.044	0.002	0.223	1.709	4.153	4.299
LIIP	4.849	0.048	0.002	0.374	3.063	4.765	4.971
LGoldPrice	7.156	0.048	0.002	1.097	4.990	7.083	7.313
LCPI	4.924	0.030	0.001	−0.350	2.267	4.870	4.977
REPO	6.145	0.254	0.064	−0.589	3.684	5.400	6.500

Notes: LNiftyAuto - Nifty Auto index, LCRUDE - crude price, LER - exchange rate, LIIP - index of industrial production, LGoldPrice - gold price, LCPI - consumer price index, REPO - repo rate. The prefix L indicates that the log form of the corresponding variable is taken.
Source: authors' compilation.

Table 3 ADF test results

	Variables	Level form	First difference
1	LNiftyAuto	−1.243(0.9014)	−6.121(0.0000)
2	LCRUDE	−1.173(0.9159)	−3.905(0.0119)
3	REPO	0.234(0.9960)	−5.232(0.0001)
4	LER	−2.381(0.3895)	−3.531(0.0362)
5	LIIP	−7.039(0.0000)	−14.325(0.0000)
6	LCPI	−1.658(0.7689)	−3.447(0.0454)
7	LGoldPrice	−0.472(0.9845)	−4.017 (0.0083)

Note: The numbers provided in parenthesis are p-values. Source: Author's compilation

From the above results it is seen that variables like LNiftyAuto, LCRUDE, REPO, LER, LCPI and LGoldPrice are integrated of order 1(i.e., stationary at first difference) whereas LIIP is integrated of order 0 (i.e., stationary at level). Hence it is ascertained that there is no problem in going forward with the ARDL tests. The PP test results also confirmed the same but are not provided here to ensure brevity.

4.3 Ardl Cointegration Test

Having established that the data series under study are a combination of I(0) and I(1) in the previous sections, now we proceed to analyze the co-integrating relationship using ARDL bounds testing approach (Adeleye et al. 2018). The step-by-step ARDL test results are discussed in the following sections. It is important to note that Stata reported a multicollinearity error with LCPI in the model and was automatically eliminated when ARDL regression was done.

Lag Length Selection

The optimal lag(s) selected by Akaike Information Criterion (AIC) for the variables LNiftyAuto, LCRUDE, LER, LIIP and REPO were 4, 4, 2, 4 and 1 respectively.

ARDL Bounds Test

$$H_0 : \text{ no levels relationship}$$

The bounds test results are provided in Table 4 below. As the F statistic value (7.036) is greater than the critical value for I (1) regressors, the null hypothesis is rejected. Thus, it suggests a co-integrating relationship between the variables (Pesaran et al. 2001).

ARDL and ECM Results

Table5 given below presents the long run estimates.

The long run results indicate that LER (-4.3025) and LIIP (-2.0583) have a negative impact on the NSE auto index. At the same time, CRUDE (0.958) & REPO (0.2461) seems to have a positive influence on the movement of the NSE auto index. The short run estimates along with the adjustment coefficients are given in the Table 6.

The error correction or adjustment term (i.e. the first lag of LNiftyAuto) is negative (-1.2318) and is found to be statistically significant as the p-value, $0.008 < 0.05$. In

Table 4 Bounds test results

Estimated Model: LNiftyAuto = F (LCRUDE, LER, LIIP, REPO)			
F statistic = 7.036			
Critical Values			
Significance levels			
Upper Bound I (1)	10%	5%	2.5%	1%
	3.52	4.01	4.49	5.06

Table 5 ARDL Long run results

Regressors	Coefficient	p-value	Regressors	Coefficient	p-value
Constant	39.20	0.004	LIIP	-2.0583	0.001
LCrude	0.958	0.000	REPO	0.2461	0.000
LER	-4.3025	0.000			
Regression Statistics					
No. of observations	28				
R^2	0.9465	Adjusted R^2	0.8195		

Table 6 ARDL short run results

Regressors	Coefficient	p-value	Regressors	Coefficient	p-value
Adjustment (ECM$_{t-1}$)			**ΔLiftyAuto**		
LiftyAuto L1	-1.2318	0.008	LD	−0.0668	0.724
			L2D	−0.6318	0.068
			L3D	−0.5579	0.010
ΔLCrude			**ΔLIIP**		
D1	−0.618	0.137	D1	2.2745	0.010
LD	−0.687	0.007	LD	1.6024	0.051
L2D	−0.154	0.413	L2D	0.995	0.051
L3D	−0.347	0.071	L3D	0.394	0.192
ΔLER			**ΔREPO**		
D1	1.815	0.117	D1	−0.2792	0.066
LD	2.226	0.216			

Note: L1, L2, and L3 indicate the first, second, and third lags of ΔLNiftyAuto. Δ indicates the first difference operation.

Table 7 Summary of diagnostic test results

Specification	Test statistic/p-value	Conclusion
JB(normality)	3.87/0.144	Evidence of normality
Durbin-Watson	2.67	No first order autocorrelation
Breusch-Godfrey	6.807/0.0091	Higher order auto correlation
Breusch-Pagan	0.02/0.8798	No heteroscedasticity
Arch LM	1.488/0.2226	No conditional heteroscedasticity
Ramsey's RESET	1.23/0.3901	No omitted variables
CUSUM (recursive)	0.2620/0.9479	No structural breaks

this analysis, ECMt-1 = −1.2318 indicates that 123.18% of the disequilibrium of the previous month's shocks are corrected back to the long run equilibrium in the present month via the explanatory variables in the model. The coefficient is less than −1 but lies within the dynamically stable range as it is not lower than −2 (Pesaran et al. 1999; Adeleye et al. 2018).

From the short run coefficients, it is seen that LER and LIIP has a positive influence whereas CRUDE has a negative influence on the auto index. These results are consistent with past studies (Joshi & Giri 2015; Sinha & Kohli 2015). Moreover, the first lag of CRUDE (−0.687) seems to be a significant predictor of Nifty auto index in the short run. For the first lag of CRUDE (i,e previous month value), this can be explained as, a 1% increase of last month's crude price will lead to approximately 0.687% decrease in Nifty auto index in the current month, ceteris paribus. Further, $R^2 = 0.9465$ indicates that 94.65% of the variations in Nifty Auto index is explained by the regressors in the model (Koop 2013).

Diagnostic Test Results
The diagnostic and stability tests results are given in Table 7.

5 Conclusion

The aim of the study was to analyze the association between selected macroeconomic factors such as crude price, exchange rate (USD/INR), index of industrial production, repo rate, inflation, gold price and the Nifty auto index. ARDL cointegration method was used to analyze the data.

Exchange rate and IIP were seen to be significant negative predictors of Nifty auto index in the long run. In addition, crude price and interest rates were seen to have a significant positive relationship with the index in the long run. On the other hand, in the short run, first lag of crude price revealed a significant negative association with the auto index while IIP showed a significant positive relationship. The above results have implications for investors, portfolio managers, corporations, and policy makers.

This study recommends that corporations may adopt appropriate hedging strategies to mitigate the exchange rate risk. Furthermore, they must also see whether the industry is up to date with the market trends and adopt appropriate strategies. Additionally, policy makers should monitor the current situation of the auto industry and implement appropriate macroeconomic policies (like scrappage policies, tax reduction) to foster growth.

Like with other empirical research, this study has some limitations that must be addressed in future research. The single equation model that we adopted for analyzing the association between the variables might not be suitable in such situations when there is inter-relationship among the chosen variables. Hence, future researchers should consider using simultaneous equation models. Even more, additional macroeconomic variables and longer time frames can be incorporated to have a better outlook on the association between the variables under study.

References

Adeleye, N., Osabuohien, E., Bowale, E., Matthew, O., & Oduntan, E. (2018). Financial reforms and credit growth in Nigeria: empirical insights from ARDL and ECM techniques. *Int. Rev. Appl. Econ., 32*(6), 807–820.

Alexander, R., & Al-Malkawi, H. A. N. (2022). On the relationship between macroeconomic factors and S&P BSE auto index: an ARDL approach. *Eur. Stud. Bus. Econ., 22*, 245–263.

Asteriou, D., & Hall, S. (2007). *Applied Econometrics*. London: Palgrave Macmillan.

Büyük, E., Otomobil, D., & Ülkedeki, Ü. (2018). Macroeconomic effect on the automobile sales in top four automobile production countries. *Kocaeli Üniversitesi Sosyal Bilimler Dergisi, 35*, 139–161.

Gaspareniene, L., & Remeikiene, R. (2014). Evaluation of the factors that influence the eu automobile industry during the period of financial crisis. *Medit. J. Social Sci., 5*(27), 1735–1740.

Gujarati, D. (2003). *Basic Econometrics*. New York: McGraw-Hill.

Hindustan Times. (2019). India's auto sales dip at fastest pace in nearly 2 decades. [Online]. [Accessed 14 August 2019].: https://www.hindustantimes.com/business-news/india-s-passen ger-vehicle-sales-drop-at-steepest-pace-in-nearly-two-decades/story-xhpS4xNSe69JvdjxER YaWM.html.

Joshi, P., & Giri, A. K. (2015). Dynamic relations between macroeconomic variables and Indian stock price: an application of ARDL bounds testing approach. *Asian Econ. Finan. Rev., 5*(10), 1119–1133.

Koop, G. (2013). *Analysis of Economic Data*. Wiley Textbooks.

Mishra, P. K., Mishra, U. S., Mishra, B. R., & Mishra, P. (2010). Capital market efficiency and economic growth: the case of India. *Eur. J. Econ. Finan. Adm. Sci., 27*, 131–138.

Misra, P. (2018). An investigation of the macroeconomic factors affecting the Indian stock market. *Aust. Account. Bus. Finan. J., 12*(2), 71–86.

Mohi-u-Din, S., & Mubasher, H. M. (2013). Macroeconomic variables on stock market interactions: the Indian experience. *IOSR J. Bus. Manag., 11*(3), 15–28.

Nanda, S., & Panda, A. K. (2018). The determinants of corporate profitability: an investigation of Indian manufacturing firms. *Int. J. Emerg. Mark., 13*(1), 66–86.

Pesaran, M. H., Pesaran, M. H., Shin, Y., & Smith, R. P. (1999). Pooled mean group estimation of dynamic heterogeneous panels. *J. Am. Stat. Assoc., 94*(446), 621–634.

Pesaran, M. H., Shin, Y., & Smith, R. J. (2001). Bounds testing approaches to the analysis of level relationships. *J. Appl. Econ., 16*(3), 289–326.

Shahabuddin, S. (2009). Forecasting automobile sales. *Manag. Res. News, 32*(7), 670–682.

Sheikh, U. A., Asad, M., Ahmed, Z. & Mukhtar, U. (2020). Asymmetrical relationship between oil prices, gold prices, exchange rate, and stock prices during global financial crisis 2008: evidence from Pakistan. *Cogent Econ. Finan.* Cogent 8(1).

Srivastava, A. (2010). Relevance of macro economic factors for the indian stock market. *Decision, 37*(3), 69–89.

Tambade, H., Singh, R. K., & Modgil, S. (2019). Identification and evaluation of determinants of competitiveness in the Indian auto-component industry. *Benchmarking, 26*(3), 922–950.

Tripathi, V., & Seth, R. (2014). Stock market performance and macroeconomic factors: the study of Indian equity market. *Global Bus. Rev., 15*(2), 291–316.

Ensuring the Fourth Industrial Revolution's Connectivity Requirements in the UAE

Khalid Alawadhi and Khalid Almarri

Abstract The international community has recognized the critical role of information and communication technology (ICT) in the advancement of humanity. ICT is a critical component of each of the United Nations' (UN) 17 Sustainable Development Goals (SDGs) (ITU 2021). Nonetheless, 48% of the world's population remains unconnected to the internet (Henri 2020, p. 2), resulting in what is referred to as the digital divide. The information and communications technology industry is making strenuous efforts to provide diverse connectivity options in metropolitan areas. However, little effort is being taken to guarantee that underserved markets are connected (Sheikh and Halima 2020). In the UAE, we identify underserved markets and classify them as rural villages, remote agricultural/industrial areas, and vehicles in motion.

By examining the present ICT market, we can discover a possible solution to this problem through the use of cutting-edge satellite communications technology. Numerous initiatives involving the use of Non-Geostationary Orbit (NGSO) satellites to deliver broadband access began to emerge. The primary benefit of NGSO satellites over Geostationary Orbit (GSO) satellites is their proximity to the Earth's surface (NGSO is approximately 2000 km above sea level, while GSO is at an elevation of around 36,000 km above sea level). The objective is to perform market research and analyze the connectivity solutions available today. Finally, we will be able to choose the option that is expected to resolve the issue.

Keywords Information and communications technology · ICT · Digital divide · Satellite connectivity · NGSO satellites · Fourth industrial revolution · 4IR

K. Alawadhi (✉) · K. Almarri
The British University in Dubai, Dubai, UAE
e-mail: khalid.alawadhi@outlook.com

© The Author(s) 2023
K. Al Marri et al. (eds.), *BUiD Doctoral Research Conference 2022*,
Lecture Notes in Civil Engineering 320,
https://doi.org/10.1007/978-3-031-27462-6_3

1 Introduction

1.1 Connectivity Expectations

Humanity continues to make major strides in the field of information technology. The promise of better communication speeds and increased bandwidth has accelerated the development of innovative technical solutions like as artificial intelligence, augmented reality apps, 3D printing, remote surgical operations, and different smart city applications. These breakthroughs, together referred to as the Fourth Industrial Revolution (4IR), are projected to cause a paradigm shift in living (Schwab 2016; Sheikh and Halima 2020). By introducing such breakthroughs, humanity's reliance on connectivity would be increased as well. Connectivity service providers are striving to introduce connectivity innovations that would meet future customer requirements (Yarali 2022).

The International Telecommunications Union (ITU), the UN specialized organization for information and communication technologies, is actively debating the issue of the digital divide: countries that are much less connected to the internet than the rest of the world. The digital gap is a serious issue that threatens the global development and diffusion of the 4IR. ITU member states are seriously considering fixing this issue in partnership with the ICT industry.

1.2 The Regulatory Framework for ICT in the UAE

ICT is a critical component of any community. Additionally, it is understood that providing ICT services is a very profitable endeavor, and leaving the business unchecked could result in chaos, lowering the quality of given services. As a result, governments typically establish a regulatory agency to oversee the information technology sector, with the goal of facilitating the supply of important ICT services to society. In the UAE, this function is performed by the Telecommunications and Digital Government Regulatory Authority (TDRA). According to the UAE's Telecom Law, every firm wishing to provide ICT services in the UAE must get a TDRA-issued telecoms license. These "licensees" must adhere to severe terms and conditions in order to ensure service quality and to provide services to the majority of the country's population. Simultaneously, the TDRA plays a critical role in safeguarding licensee interests by promoting "balanced competition" in the sector. Too many ICT solution providers would erode the licensee's market share, ultimately leading in service degradation.

1.3   The Underserved Market

The UAE recognized the critical nature of ensuring dependable connection in order to prepare for the 4IR applications' integration into society. Providing such creative solutions to a subset of the UAE population would be socially unacceptable. It would also undermine the objective of the UAE Government's Centennial Plan 2071 to embrace a high-quality lifestyle for the UAE society (UAE Portal 2022). The UAE's current internet connectivity is considered to be among the greatest in the world. In January 2021, the UAE ranked first globally in terms of internet speed, with an average of 183 megabits per second (Ookla 2021).

There are always people who are not well connected in every country (Henri 2020). Although the average internet speed in the UAE is considered the best in the world, some underserved markets are still identified. Generally, the ICT industry makes many efforts to provide connectivity solutions in urban areas in anticipation of the 4IR. However, not many efforts are being made to ensure the connectivity of underserved markets (Sheikh and Halima 2020). Most of the time, companies are driven by profitability. If the decision on service provisioning was left entirely to the private sector, ICT services would have been provided only to urban areas: regions with a dense population that ensure a good Return of Investment (ROI) for the laid ICT infrastructure. Therefore, regulators such as the TDRA ensure that allowing entities to provide ICT services comes with conditions to ensure good service coverage. However, regulators cannot force any service providers to provide services. Because of investment costs (Lin et al. 2021), there are always areas with a minor population that would be very difficult to reach with ICT infrastructure, such as The Empty Quarter Desert in the case of the UAE. We define these underserved markets as locations where the broadband connectivity speeds do not yet reach the expected minimum requirements for 4IR applications to be provided. These underserved markets are categorized into three categories:

- Rural villages.
- Remote agricultural/industrial locations.
- Vehicles in motion.

The first category of underserved markets is the rural villages. These villages are either far away from urban areas or with rugged geographical terrain, making it unfeasible for licensees to provide high-speed internet connectivity. Some of these villages have connectivity to cater to the license conditions of the licensees. However, this connectivity (usually a 3G network) would not suffice for the future connectivity expectations of the 4IR. The second category of underserved markets is the agricultural fields and industrial locations, such as oil rigs. Usually, these areas do not have a considerable population. No one resides in these locations, and the only people in the area are the workers in these fields. However, the expectations of 4IR demand for availability of good connectivity in these locations. The third category of underserved markets is the vehicles in motion. Connectivity is not a concern for vehicles within urban areas, such as cars, public buses, and metros within cities. For vehicles that move across Country borders (transportation trucks and trains), ships, and planes, minimum connectivity is currently ensured by international regulatory bodies to guarantee the safety of life applications. However, this minimum connectivity will not suffice for the provisioning of 4IR solutions. The vehicles themselves will need sufficient connectivity for IoT applications. Their passengers will also need the required connectivity to ensure the continuation of 4IR applications, which are expected to be an essential part of future life. Besides the crew of these vehicles, the primarily affected users in this category are travelers onboard planes, trains, and cruise ships. These passengers expect to continue their ordinary life that is (expectedly) filled with 4IR use cases while in transit.

1.4 Current Connectivity

In the UAE, standard internet connectivity is provided by licensed Internet Service Providers (ISPs). ISPs usually offer broadband services to fixed locations using fiber cables laid underground (fiber to home). For areas not yet served by fiber cables, ISPs provide other wireless connectivity technologies, such as Fixed LTE. These terrestrial wireless connectivity solutions are considered suitable for the expected future connectivity requirements. Also, mobile broadband is being provided by the existing ISPs using mobile technologies. Current mobile broadband solutions are considered sufficient to support today's connectivity requirements. For example, 4G connectivity would be sufficient to stream live videos while in motion (Zhang et al. 2018). However, it would not entertain the requirements of the expected future technologies, such as autonomous vehicles. Therefore, connectivity advancements, such as 5G, are being introduced to cater for the future connectivity requirements (Mumtaz et al. 2020). For the identified underserved markets, these futuristic connectivity solutions are not currently provided by the existing ISPs.

Another connectivity solution exists for the underserved markets by licensed satellite service providers. Internet connectivity can be provided currently by satellites that are located in the GSO. However, these connectivity solutions suffer from propagation delays (Lin et al. 2021), and connectivity speeds are less than the average mobile broadband provided by the existing ISPs and much less than the minimum required connectivity to provide 4IR applications.

2 Literature Review

To tackle the national digital divide problem, we attempt to look into available connectivity solutions by reviewing the literature. The following possibilities were identified.

2.1 Terrestrial Backhauling

A possible solution to the stated issue would be to enable the establishment of infrastructure by existing ISPs. Typically, the reason why present ISPs are unable to provide critical services to underserved communities is the high cost of establishing the necessary infrastructure. At the moment, connectivity is provided via four distinct forms of terrestrial infrastructure (Lambrechts and Sinha 2019):

– a mobile network that adheres to globally recognized connection standards, such as 4G and, more recently, 5G
– copper cables
– fiber optic cables
– fixed wireless access

However, new technologies have recently emerged for terrestrial infrastructures, such as Long Range (LORA), Narrow Band IoT (NB-IoT), Multi-hop Wi-Fi technology, Unmanned Aerial Vehicles (UAVs), Light Fidelity (Li-Fi), and Visible lightwave and Infrared (IR) (Challita and Saad 2017; Lambrechts and Sinha 2019; Sheikh and Halima 2020). Using combinations of these advancements could help ISPs provide the required backhauling for underserved markets for a low cost. Lambrechts and Sinha (2019) demonstrate the advantages and disadvantages of some of the infrastructure solutions (including conventional satellite connectivity) in the following comparative table (Table 1):

Taking such an approach to solving the national digital divide problem would be the easiest from the regulatory perspective. The problem would be resolved through investments by the existing ISPs with no intervention required from the regulator.

2.2 Satellite Connectivity by NGSO Satellites

The utilization of the latest developments in satellite communications could be another solution to the identified problem. Many initiatives started to emerge involving the use of NGSO satellites to provide broadband connectivity. The most significant advantage of NGSO satellites over GSO satellites is that they are much closer to the Earth's surface: NGSO is 2000 km above sea level, while GSO is at an elevation of around 36,000 km above sea level. Figure 1 shows the different orbits available for satellites around the Earth and their typical utilization. Coupled with the latest advancements in satellite technologies, low Earth orbit allows for much faster communications with more reliability and higher capacity, as propagation delay is dramatically less than the propagation delay for a GSO satellite (Henri 2020; Lin et al. 2021). Initially, two-way satellite communications could be provided by GSO satellite only to fixed locations on Earth, such as services provided by Intelsat. Later development in satellite communications allowed connectivity between GSO satellites and mobile terminals, with low speed and data rates. Now, we are on the verge

Table 1 Advantages and disadvantages of infrastructure solutions

Mobile (GSM) networks—advantages	Mobile (GSM)—disadvantages
High level of network access mobility	Broadband network access is costly
Relatively good service reliability	Cellular network access in rural areas are limited
Cost-effective to gain basic access to networks	Network speed varies with external factors such as weather and foliage
DSL/cable—advantages	DSL/cable—disadvantages
High level of network availability—especially in urban areas	Performance degradation a strong function of distance to the core network
Low infrastructure cost for access to homes and businesses	Speed limitations hinders future-proofing infrastructures
High service reliability and quick turnaround times for network maintenance and repairs	Copper cables prone to long-term damage and corrosion
Fiber—advantages	Fiber—disadvantages
Very high bandwidth capabilities at low latency	High cost for individual users
No susceptibility to electromagnetic interference	Glass wire cables prone to physical damage
Lightweight cables, service reliability and long lifespan	
Fixed wireless—advantages	Fixed wireless—disadvantages
Cost-effective to gain basic access to networks	Network speed varies with external factors such as weather and foliage
Standard protocols such as WiMax already established	Network reliability and transmission speeds dependent on line-of-sight access
Potentially deliver transfer speeds of several 10's of Mbps	High latency typically associated with fixed wireless
Satellite—advantages	Satellite—disadvantages
Accessible from virtually anywhere in the world	Network speed varies with external factors such as weather and foliage
High service reliability	High latency typically associated with connections
Li-Fi—advantages	Li-Fi—disadvantages
Low investment cost per household or business	Network reliability and transmission speeds dependent on line-of-sight access
Very high bandwidth transmissions	Always-on lights required for network access
Multiple users on single connection without degradation	Susceptible to environmental interference from other/ambient lights

Source: (Lambrechts & Sinha, 2019, *p. 12*)

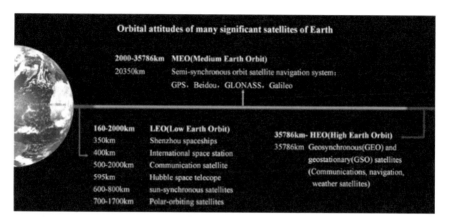

Fig. 1 Distance of different orbits from Earth and their typical utilization **Source:** *(Lu et al. 2019, p. 93,474).*

of the most significant development in satellite communications so far, with capabilities comparable to terrestrial radiocommunication. If we could utilize this innovation by providing such connectivity within the UAE, it would be a solution to provide the required connectivity to the underserved market. Along with the advancements that improved connectivity via satellite came the dramatic reduction in the cost of building and launching small NGSO satellites. As a result, we now have satellite connectivity that could be provided anywhere on Earth with comparable capabilities to terrestrial networks and comparable prices.

Two solutions could be considered for the utilization of satellite connectivity to resolve the national digital divide issue. The first solution is to provide satellite backhauling to existing service providers that are licensed in the UAE. The second solution could be to provide services to end-user customers directly from satellite operators.

2.2.1 Satellite Backhauling by NGSO Satellites

Satellite backhauling is already being considered by many terrestrial telecommunication service providers around the world. Satellite backhauling is regarded as a major part of the business model for many satellite operators. Typically, these backhauling solutions allow service providers to provide their services in remote locations, which would be cheaper than laying down the required cable infrastructure for the service. So far, satellite backhauling was through GSO satellites, and the quality of services provided through this backhauling would not be comparable to standard terrestrial connectivity.

With NGSO satellite connectivity, competitive services could be provided to the identified underserved markets. Also, there will not be a requirement to amend the existing regulatory framework by the TDRA, as the existing ISPs will provide the services to the public. Some NGSO satellite operators, such as OneWeb (Henri 2020), include the possibility of providing backhauling services to existing ISPs. One disadvantage in this arrangement is that the NGSO satellite operators would usually prefer providing the services directly to end-users, as their connectivity capabilities are similar to terrestrial networks. Direct service to end-users would be more profitable than leasing some satellite capacity to terrestrial service providers. Hence, there is a probability of not reaching an acceptable agreement between the two parties, or the commercial arrangement might not be feasible to end-users.

2.2.2 End-User Service Provisioning by NGSO Satellites

As satellite connectivity became more attractive through NGSO satellites, it opened the opportunity for NGSO satellite operators to directly provide connectivity services to end-users. Some NGSO satellite operators are taking this approach in providing their services globally, such as Starlink (Starlink 2021). The satellite industry has recently realized this solution, and efforts have been made recently to align satellite connectivity with the terrestrial broadband connectivity standards. This solution is also recognized as the only connectivity possibility for the third category of underserved markets, which is the vehicles in motion (Lin et al. 2021).

Though it is more convenient and commercially attractive for the satellite operator to adopt this approach, it has many challenges. First of all, similar to the UAE, there is a requirement to obtain a telecommunications service provisioning license in many countries worldwide, in addition to satellite signal landing rights. The satellite operator has to approach every regulatory body in all countries in which they wish to provide their services. There is also a significant probability that they will not be able to receive the required approvals. Another challenge is the compliance with various regulatory requirements in every country, which could create a technical challenge for the NGSO satellite operator. It might not be possible to control the satellite transmissions differently for signals received from different adjacent countries.

Starlink is one of the latest promising initiatives for NGSO satellite connectivity. It is a project launched by SpaceX, a pioneering entity in the space industry, owned by the entrepreneur Elon Musk. Starlink is promising internet connectivity with speeds ranging from 50 to 150 Mbps (Starlink 2021). As the company plans to launch thousands of satellites to provide connectivity, the connectivity speed is expected to increase with more satellites launched. This expectation is recognized as technically correct, but we will not go into these technical details within this paper. The service price will be $99 per month, in addition to an initial investment of $499 for the receiving kit (Sheetz 2020). Table 2 illustrates the differences between the three identified solutions in this study.

Table 2 Comparison between the available solutions

	Terrestrial backhauling	NGSO satellite backhauling	NGSO satellite services to end-users
The requirement for regulatory amendments	Not required	Not required	new telecom licenses are required for potential NGSO satellite service providers
Service prices	Similar to existing prices	Similar to existing prices	Comparable to existing prices
Connectivity quality	Similar to urban connectivity	Not similar to urban connectivity, but sufficient	Not similar to urban connectivity, but sufficient. Expected to increase
Service to underserved markets	no possibility of service to vehicles in motion	no possibility of service to vehicles in motion	All underserved market categories
Preference by existing ISP	Yes	Yes	No
Preference by NGSO satellite operators	No	No	Yes

3 Conclusion

ICT advancements have enabled the development of novel methods for connecting previously unconnected places. As the global society is worried about the digital divide phenomenon, the TDRA, as the UAE's regulator of the ICT sector, should ensure that no location inside the UAE is unconnected or poorly connected. Addressing this national digital divide is crucial for the UAE's 4IR applications and use cases to be successfully deployed. We examined various solutions available on the market now in this study. We compared the various solutions. Additional research is required to determine the financial impact of applying these solutions on the ICT market and existing ISPs.

After comparing the various solutions identified, we advocate delivering NGSO satellite connectivity to end customers in the identified underserved markets. This method is widely considered as the sole viable option for providing connectivity to moving cars. Additionally, the service is supplied at a price comparable to what is already offered by established ISPs. The network speed currently available is expected to be sufficient for 4IR applications. Additionally, the speed of connectivity is predicted to rise as the number of satellites launched increases.

Until now, the UAE has awarded licenses to particular ISPs in order to limit the number of service providers and maintain market stability. The TDRA may contemplate service offering to certain locations/markets within the UAE. For instance, a license could be granted to deliver Broadband Services within airplanes in the UAE or Internet of Things services in UAE oilfields. Numerous points need to be resolved before pitching this proposal, including the market share of existing operators in the targeted markets, the influence on revenue generation for existing operators, and the predicted service enhancement for the targeted market.

It is critical to note that we are not limited to a single solution, since multiple solutions could be adopted to ensure redundancy and the availability of alternatives (Lambrechts and Sinha 2019). Additionally, to this method, a combination of terrestrial backhauling systems could be used. This, however, will depend on operators' willingness to invest in these infrastructure solutions based on their expected return on investment.

References

Challita, U., & Saad, W. (2017) Network formation in the sky: unmanned aerial vehicles for multi-hop wireless backhauling. In: *2017 IEEE Global Communications Conference, GLOBECOM 2017*, pp. 1–6

Henri, Y. (2020) The oneweb satellite system'. In: *Handbook of Small Satellites*. Springer Nature Switzerland AG 2020, pp. 1091–1100

ITU (2021) *ICTs for a Sustainable World #ICT4SDG*. Accessed 8 June 2021. https://www.itu.int/en/sustainable-world/Pages/default.aspx.

Lambrechts, W., & Sinha, S. (2019) Last mile internet access for emerging economies. In: *Lecture Notes in Networks and Systems*. Heidelberg: Springer.

Lin, X., et al. (2021) 5G new radio evolution meets satellite communications: opportunities, challenges, and solutions. In: *5G and Beyond*. Springer, pp. 517–531.

Lu, Y., Shao, Q., Yue, H., & Yang, F. (2019). A review of the space environment effects on spacecraft in different orbits. *IEEE Access, 7*, 93473–93488.

Mumtaz, T., Muhammad, S., Aslam, M. I., & Mohammad, N. (2020). Dual connectivity-based mobility management and data split mechanism in 4G/5G cellular networks. *IEEE Access, 8*, 86495–86509.

Ookla. (2021). *Speedtest Global Index*. Accessed 9 Aug 2021. https://www.speedtest.net/global-index.

Schwab, K. (2016) *The Fourth Industrial Revolution: what it means, how to respond*. https://www.weforum.org/agenda/2016/01/the-fourth-industrial-revolution-what-it-means-and-how-to-respond/.

Sheetz, M. (2020) *SpaceX prices Starlink satellite internet service at $99 per month, according to e-mail. CNBC*. Accessed 4 Aug 2022. https://www.cnbc.com/2020/10/27/spacex-starlink-service-priced-at-99-a-month-public-beta-test-begins.html.

Sheikh, S.M. & Halima, N.B. (2020) Building backhaul networks for rural area connectivity towards the fourth industrial revolution. In: *2020 3rd International Conference on Computer Applications & Information Security (ICCAIS)*. IEEE, pp. 1–4.

Starlink. (2021) *High-speed low latency broadband internet*. Accessed 8 June 2021. https://www.starlink.com/.

UAE Portal (2022) *UAE Centennial 2071*. Accessed 4 Aug 2022. https://u.ae/en/about-the-uae/strategies-initiatives-and-awards/federal-governments-strategies-and-plans/uae-centennial-2071.

Yarali, A. (2022). *Intelligent Connectivity AI, IoT, and 5G* (1st ed.). Hooken: John Wiley & Sons Ltd.

Zhang, J., Wang, Z. J., Quan, Z., Yin, J., Chen, Y., & Guo, M. (2018). Optimizing power consumption of mobile devices for video streaming over 4G LTE networks. *Peer-to-Peer Netw. Appl., 11*(5), 1101–1114.

Enhancing Products Delivery Through the Application of Innovative Operating Model Based on Hybrid Agile Delivery Method: Case Information Communication Technologies "ICT" Service Providers

Madhad Ali Said Al Jabri

Abstract **Purpose:** the purpose of this study is to find out a cutting-edge innovative solution for enhancing ICT products delivery and proposing an implementation plan with transformation recommendations aiming higher success rate.

Methodology: This qualitative study is done through intensive literature review followed by qualitative analysis and selection of best product delivery method using cost benefit analysis and previous successful implementations benchmark. A proposed delivery operating model was also recommended accordingly.

Findings: Based on the problem analysis and subsequent literature review, the existing delivery methodologies of ICT services providers are not fully sufficient to cater for high complex products delivery which require high integration and customer centricity. Hence, a hybrid delivery method is recommended based on previous success stories of 5 major companies with a proposed transformative operating model (framework) including a transformation plan and implementation recommendations.

Implications: Information Communication Technologies services providers can benefit from the proposed hybrid product development approach and the suggested operating model as key differentiator for agility, customer centricity, and improved time to market. It enhances collaboration, innovation, culture, and Employee Engagement.

Originality/value: Previous literature has extensively explored projects delivery models in various industries where there was minimum focus on ICT industry. This paper uncovers and contribute positively in proposing a successful hybrid agile delivery method with a transformative operating model to support higher success rate in ICT Industry to assure high-value creation, maximize return on investments with high net present value.

Keywords Products development · Project management · Agile development · Agile organizations · And digital delivery

M. A. S. Al Jabri (✉)
The British University in Dubai, Dubai, United Arab Emirates
e-mail: 21002103@student.buid.ac.ae

© The Author(s) 2023
K. Al Marri et al. (eds.), *BUiD Doctoral Research Conference 2022*,
Lecture Notes in Civil Engineering 320,
https://doi.org/10.1007/978-3-031-27462-6_4

1 Introduction

With the advancement of new digital technologies and enhancement of quality-of-life requirements, consumers and businesses are increasingly demanding for new digital services and higher customer experience. This invites ICT services providers to innovate and offer new digital services with high quality and agility in such fast-changing environment (Albers et al., 2019).

Information and Communication Technology Services Provider are known for their technology and products leadership across all decades. With the Industry 4.0 revolution, a major focus is provided to transform their culture towards innovation-driven based delivery to overcome increasing needs of digital products especially business services where delays are encountered with quality problems causing efficiency issues and user experience issues.

In order to support ICT Services Providers to fast track their innovation and products leadership, this study explore the literature to find out a cutting-edge innovative solution to be taken into consideration to transform the existing delivery method towards achieving faster time to market and maintain their technology leadership using transformative framework (Heimicke et al., 2021).

Several strategies found in the literature to accelerate the products delivery such as the need for proper staffing of product development teams, usage of focused teams, and effective portfolio management deployment. This involves prioritization and resources reallocation. Also, new digital technologies are required to accelerate the delivery through agile and lean development which are often used interchangeably (Cooper, 2021).

Several design principles are considered such as Time to Market, Speed of Delivery, Quality, and Voice of Customers. Hence, a hybrid delivery method is recommended with a proposed framework and recommendations.

2 ICT Environment Analysis

2.1 Industry 4.0 Services Scaling Up Rapidly

In alignment with the Government Strategies, the ICT Sector introduced the 5G Network with huge potential of digital solutions. These enables to meet essential needs of improving quality of life, enable corporate social responsibility, enhance customer experience, and support the digital economy.

> 5G has a huge potential of delivering a distinct product portfolio supporting the societies. It is enabling today's new use cases driving new products development with a high level of complexity such as cloud computing services, edge services, private networks, business edge solutions, IoT solutions, Safety and control solutions, security and surveillance solutions, traffic control solutions, vehicle tracking solutions, connect cars solutions, health solutions and banking solutions (Rao et al., 2018).

The above solutions and opportunities to be delivered in a fast manner need huge collaborative efforts across the full supply chain and a high level of communication among the development teams and suppliers as well as systems integrators. The existing operating model and development processes need a revamp by the ICT Company under this case study to enable continual technological leadership and faster time to market targets achievements (Albers et al., 2019).

2.2 Pandemic Increased Demand for New Services

The recent spread of Coronavirus Disease (COVID-19) in the last two years generated a huge amount of demand on new services that enable health, education, remote working environment, entertainment anywhere, new business products those enabling higher firms' efficiency and performance, logistical services, smart cities services, online services, governmental digital services. Hence, Rapid new-product creation is more vital than ever, and it highlighted instances of companies who created game-changing products at record speed (Cooper, 2021).

In Parallel, the level of competition is increasing in the marketplace as hyper-scalers such as Amazon, Google, Microsoft, others are competing strongly to serve their customers through the pandemics, as an example the sharp scale-up of Microsoft Teams, Zoom, food delivery, and other smart surveillance services and home services targeting home and office eco-system products.

Therefore, in order to respond to these market forces, the ICT services providers are striving towards enhancing their agility, re-engineering its functional processes, improving communication, and focusing on talent development as key.

2.3 Problem Identification

The high demand for digital products is driven by the increasing use of internet (Kemp, 2020) and need for smart societies to improve living standards globally which require more complex in nature products that needs high integration capabilities both by Services providers and their partners. New products also need higher compliance to the high standards and quality is a pre-request by Regulators. Most importantly, customers expectations and products specifications shall meet all legal requirements as well.

Projects, like business environments in general, become increasingly technologically complex, with a higher number of tasks and complex interrelations and inter-dependencies. Due to adaptations to unanticipated and dynamic changes in the project environment or within the project itself, revisions to the initial plan are unavoidable (Collyer et al., 2010). Furthermore, the inability to precisely identify project goals is a major reason for project teams' lack of planning upfront at an early stage of the project (Chin, 2004; Shenhar & Dvir, 2007). Moreover, according to (Williams, 2005), structural complexity, ambiguity in goal definition, and project time restrictions are also the key reasons why the conventional method is ineffective for the bulk of today's projects.

Due to the nature of the information and communication business and the evolution of the conventional and traditional products which were simple and not sophisticated, the products development teams were following a basic project management approach. These methods are not in a position to properly reflect all complexity and dynamics of today's projects (Cicmil et al., 2006; Collyer et al., 2010; Williams, 2005). Therefore, ICT Service Providers under the study are facing challenges such as coping with the increasing complexities of products development, vendors alignment, demands alignment/prioritization, teams' communication issues and hence delays of product launches including quality issues which impacts market share, revenue and market leadership.

2.4   Identification and Evaluation of Alternative Solutions

The problem and issues described in the previous section necessitates to find out a suitable approach to transform the ICT Services Providers Sector towards delivery agility and an appropriate selection of project delivery approaches is critical to support faster time to market and product experience goals. Hence, the following elaborates on possible innovative solutions.

2.5   New Products Delivery Approaches

Based on the literature review, several cutting-edge innovative solutions which could be recommended methods such as: Customer-Centric Delivery, Iterative and Spiral Product Development, Idea to Launch Stage & Gating System, Agile Project Management and Hybrid Delivery Model. These alternatives are elaborated and evaluated in the following sections.

Customer-Centric Delivery

Understanding customer needs is a fundamental pre-request for organizational marketing-oriented firms where customers' requirements are thought of in-depth and considered as essential success criteria for new products, (Cooper 2018). By doing so, the voice of the customer is built in the design of the product at an early stage and calibrated across the full development life cycle of the product. This is an area of challenge for most organizations while delivering their products. The below Fig. 1 is proposed by (Cooper 2016a) which is highly recommended in delivering innovative products with higher complexity.

Iterative and Spiral Product Development Method

In the innovative studies of Cooper in product development suggested the concept of build, test, feedback and revise which is enabling continuous improvement of the intended product where the features and scope definition in lower maturity and this is the case today where the complexity is high and ICT products require a lot of efforts and integration over digital native architecture design (Cooper, 2014).

In this approach, the agility increases and the adaptability to the changes in the product's specifications across the various front-end stages of the innovation process (ideation, conceptual, design) becomes doable and achievable.

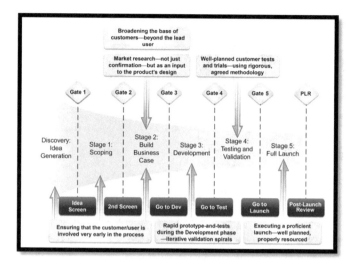

Fig. 1 Customers' requirements and involvement across the full project life cycle and stages-gates Method (Cooper, 2016a)

The cycle of this approach starts with the building phase where the product is built to demonstrate to the customer or his/her representative from the commercial teams via a prototype and other similar models. Then the product is continuously tested with the customer. Users' feedback is obtained and provided instantaneously to the design team for immediate improvement and finally, the value proposition is revised and moved on to a new iteration cycle as indicated in Fig. 2 below.

This spiral process is cost-efficient because it allows an early trial and error stage, enabling experimentation and testing the acceptance before detailing and delivering the product which

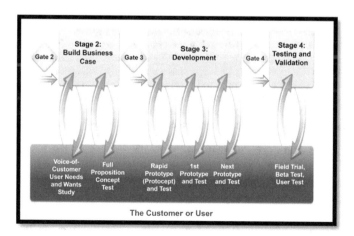

Fig. 2 Prevention of time and resources wastage through real-time feedback iterative process (Cooper, 2016b)

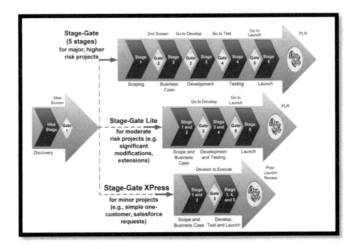

Fig. 3 Stage-gate is context-based and scalable—one size does not fit all. *Source*, (Cooper 2016b)

thus enhance the success rate. Therefore, The ICT services providers sector who deliver products to B2B customers, this process is fit for purpose especially where customers will be able to validate high expensive development and enables faster time to market accordingly.

Idea To Launch – The Stage & Gating System Method

The stage and Gating approach is widely adopted in recent years which follows an organized step starting from ideation till the launch of the product, (Cooper 2018). Each product pass through this journey where value is assured and governed to ensure that all criteria set are met and enable the project teams to move on from one stage to another one. This approach is designed for three types of projects based on risk level as per Fig. 3.

This staging system enhanced and reached a new level of maturity over time with new practices, (Cooper 2016b) which match the ICT sector products projects' agility requirements. For instance, low-medium risk projects stage and gate process suites technology platform development, ICT typical projects. Lean process across full cycle the product development where unnecessary steps are removed and value is assured. It is also considered as an open innovation adaptable system (Grölund et al. 2010). It enables end to end integration and adaptive and iterative (Spiral approach) and featured with autonomous development across the full cycle (idea-launch) using agile software development techniques.

Agile Project Management Method

"Agility - based on the system triple approach - is the ability of an operating system to continuously check and question the validity of a project plan with regard to the planning stability of the elements of the system triad and, in the case of an unplanned information constellation, to adapt the sequence of synthesis and analysis activities according to the situation and requirements, thereby specifically increasing the benefits for customers, users, and providers", (Albers et al., 2019).

The process consists of small iterations where partial products features (marketable) is delivered. Proactive customer testing is key and feedback is enabled instantaneously which helps fast detection of issues and quick changes implementation, (Aguanno, 2004). According to (Benediktsson & Dalcher, 2005), project scope can be adjusted by up to 30% during each iteration stage. Table 1 shows a comparison between conventional and agile approaches.

Table 1 Comparison between traditional and agile project management implementation methods

	Characteristics	Traditional project management approach	Agile project management approach
1	Environment	High Stability and can be predicted	High changes, hostile, and can't be predicted
2	Project Size	Mega and bigger	Smaller in size
3	Requirements and Specifications	Big adjustments aren't expected, as it's been stated upfront	Changes were encouraged after a rough first definition and subsequent development in the form of user stories
4	Planning	For the entire endeavor, detailed early preparation is required	Iterative and adaptive planning, with upfront thorough preparation for only one iteration
5	Development Approach	Stages of development must be completed in the order specified in the initial project plan	Short cycles of development, repetitive and development in drops
6	Team Work	Bigger teams, higher levels of expertise, higher levels of reporting, and clearly defined tasks	Smaller groups, generalizing experts, self-management, ring-fence teams, under one roof, open dialogues, and continuous communication on a day interval
7	Management Style	Strict governance	Cooperation, collaboration, and accountable
8	Customer Collaboration	Engagement at the start (chartering stage)	Embedded within the team with continuous opinions flows and requirements reflection
9	Communication	official	Casual
11	Quality Control	workflow-based, in depths planning, intensive governance, testing at the final stage	Human-centric, systematic control of specifications and designs, regular testing
12	Goals	Optimizing style, futuristic	agile, subject for changes, able to respond quickly to needs, fast execution with higher expectations
13	Project Success	Based on time, cost, quality parameters driven	Based on time, cost, quality with owners requirements fulfilment and delightful

Source (Žužek et al. 2021)

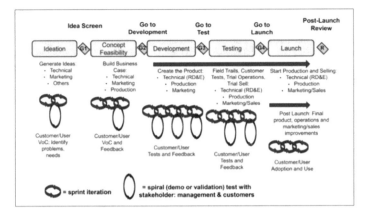

Fig. 4 Typical agile built-in staging system (hybrid model), (Cooper 2018)

Hybrid Model Approach

As ICT companies are very much known and oriented in developing B2B Products where the risk is increasing, building agile (popular in software development field) into the stage and gating system is essential. The specifications are developed through collaborative and self-managed project teams. This is also known as adaptive in-nature and customer centric planning process with stepwise development including higher rate of response to changes, (Beck et al., 2001) as shown in Fig. 4 below.

The benefits of the hybrid model are: Speed of delivery, dedicated resources, high collaboration and communication including capturing customer voice during the process.

As per (Cooper & Sommer, 2018), early adopters such as Chamberlain, Danfoss, GE (General Electric), Honeywell and LEGO Group to this agile–stage-gate model showed high success rates. They were able to have higher flexibility, productivity, communication, coordination, prioritization/focus, and teams' motivation. Hence, they have improved Time to Market by 30%. Therefore, ICT Services providers are recommended to use this model with a project selection matrix (risk based).

A Cost-Benefit analysis was carried out to select the best approach which is summarized in Fig. 5 below.

2.6 Findings and Recommendations

Several key success factors which influence the product's agile delivery (Cooper 2019) such as superior, unique, customer-focused, due diligence prior to starting the project, well defined and agreed project, use of iterative model while teams are engaging the customer upfront, including marketing launch plan were reviewed.

Based on (Heimicke et al. 2021) study, a systematic transformation is important in the current business environment. The implementation of agility in the delivery processes is a highly complex matter. The adoption of agile methods needs a certain framework to be built by the

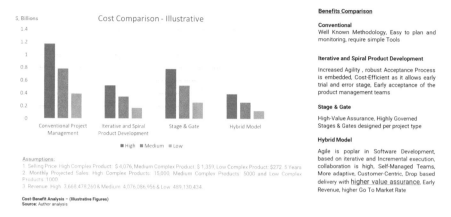

Fig. 5 Cost-benefit analysis

organization. Hence, based on the study and the above analysis, ICT Services Providers is recommended to follow the hybrid delivery approach which fits this industry and develop the below proposed operating model as illustrated in Fig. 6 and described below.

The proposed operating (framework) is designed considering strategy, agile delivery and business impacts accordingly. It starts with obtaining customer needs followed by corporate strategy development and technology strategy planning (which avoids unfamiliar technologies to minimize failures). Th same inputs to a centre of excellence which assess and prioritize requirements following stage and gate process (to assure value) then push qualified demand to the front end and concept engineering who use design thinking method leading to agile teams' execution (iterative agile development approach, (Cooper 2019)). The agile team (core and project cross functional teams) communicate tasks prioritization and results (Time

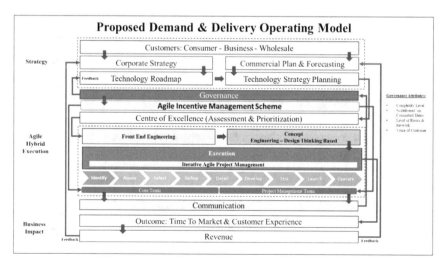

Fig. 6 Proposed new operating model

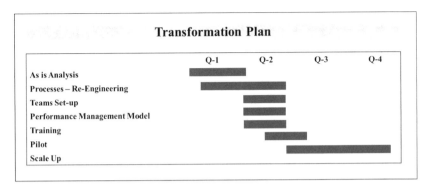

Fig. 7 Proposed transformation plan

to Market, Customer Experience) which are governed and linked to an incentive scheme that enhances collaboration and agility. Based on revenue outcomes, feedback is provided to improve both the corporate strategy and commercial plans to maintain market leadership and growth.

In order to maximize the success rate of this transformation, the following recommendations and implementation plan is proposed as shown in Fig. 7.

1. Prepare a clear development plan with full details of the communicational plan (branding), financial plan, resources plan, implementation timeline, products design strategy, processes development plan, piloting plan, scaling up, performance targets and performance reviews schedule.
2. Develop Business Case for the proposed operating model and the method selected then obtain formal management approval to pitch this transformation
3. Conduct a kick off meeting to engage stalkeholders upfront covering contracts, finance, procurement, operations, human resources, commercias and technology teams.
4. Engage the technology partners and offer the methodology as shared methodology for alignment especially ofr complex products
5. Obtain feedback from the commercial teams and customers and re-iterate the processes for continuous improvement

In conclusion, based on the research and in order to enhance the delivery of new products according to the customer's expectations, the following is recommended:

1. Implement the proposed Operating Model embracing the hybrid agile development approach supported by a prioritization process, project selection guidelines based on complexity and risk level.
2. Conduct a Pilot for the above proposed Model with clear success criteria driven by Time to Market and Voice of Customer and then scale-up
3. Communicate products priorities to raise the sense of urgency on a regular basis for teams' alignments cross-departmental

4. Establish a Performance Governance with advisory board considering measurements such as Time to Market, Quality, Voice of Customer, Service Level Agreements, etc. with shared Key Performance Indicators to enable teams' collaboration
5. Establish continuous processes optimization based on customer feedback and delivery performance
6. Deploy a new Competency Scheme to enhance product quality, learning and agility while focusing on customer engagement and education

References

Albers, A., Hirschter, T., Fahl, J., Reinemann, J., Spadinger, M., Hünemeyer, S., & Heimicke, J. (2019). Identification of indicators for the selection of agile, sequential and hybrid approaches in product development. *Procedia CIRP, 84*, 838–847.

Aguanno, K. (2004). *Managing Agile Projects*. Lakefield: Multi-Media Publications Inc.

Beck, K., et al., Principles behind the Agile Manifesto. Manifesto for Agile Software Development (2001).http://www.agilemanifesto.org/principles.html. Accessed Jan 2018

Benediktsson, O., & Dalcher, D. (2005). Estimating size in incremental software development projects. *IEE Proceedings – Software, 152*(6), 253–259.

Cicmil, S., Williams, T., Thomas, J., & Hodgson, D. (2006). Rethinking project management: researching the actuality of projects. *International Journal of Project Management, 24*(8), 675–686.

Chin, G. (2004). *Agile project management: How to succeed in the face of changing project requirements*. New York: AMACOM.

Collyer, S., Warren, C., Hemsley, B., & Stevens, C. (2010). Aim, fire, aim – project planning styles in dynamic environments. *Project Management Journal 41*(4), 108–121.

Cooper, R. G. (2014). What's next? After stage-gate. *Research-Technology Management, 157*(1), 20–31.

Cooper, R. G., & Sommer, A. F. (2016a). The agile–stage-gate hybrid model: a promising new approach and a new research opportunity. *Journal of Product Innovation Management, 33*(5), 513–526.

Cooper, R. G., & Sommer, A. F. (2016b). Agile–stage-gate: new idea-to-launch method for manufactured new products is faster, more responsive. *Industrial Marketing Management, 59*, 167–180.

Cooper, R. G., & Sommer, A. F. (2018). Agile-stage-gate for manufacturers – changing the way new products are developed. *Research-Technology Management, 61*(2), 17–26.

Cooper, R. G. (2021). Accelerating innovation: some lessons from the pandemic. *Journal of Product Innovation Management, 38*(2), 221–232.

Grölund, J., Rönneberg, D., & Frishammar, J. (2010). Open innovation and the stage-gate process: a revised model for new product development. *California Management Review, 5*(3), 106–131.

Heimicke, J., Dühr, K., Krüger, M., Ng, G. L., & Albers, A. (2021a). A framework for generating agile methods for product development. *Procedia CIRP, 100*, 786–791.

Rao, S. K., & Prasad, R. (2018). Impact of 5G technologies on industry 4.0. *Wireless Personal Communications, 100*, 145–159.

Shenhar, A. J., & Dvir, D. (2007). *Reinventing Project Management: The Diamond Approach to Successful Growth and Innovation*. Boston: Harvard Business Press.

Kemp, S., Digital 2020-July Global Statshot Report, Datareportal.com, no. 3 (2020). https://datareportal.com/reports/digital-2020-july-global-statshot

Williams, T. (2005). Assessing and moving on from the dominant project management discourse in the light of project overruns. *IEEE Transactions on Engineering Management, 52*(4), 497–508.

Žužek, T., Gosar, Ž, Kušar, J., & Berlec, T. (2021b). A new product development model for SMEs: introducing agility to the plan-driven concurrent product development approach. *Sustainability, 13*(21), 12159.

Successful Practices of Leadership on School Improvement: A Case Study in a Private School in Dubai

Ahmed ElSayed Abouelanein and Mohamed Hossni

Abstract This paper aimed to investigate successful practices of school leadership that lead to school improvement. To achieve this, the paper adopted the mixed-method approach and utilized two instruments: The first instrument is teachers' and leadership members' perceptions of school improvement questionnaire to collect quantitative data, the second instrument is a semi-structured interview with the school principal for the qualitative data. The analysis of both collected data led to a conclusion that school principal careful and professional practices have a significant impact on the overall school improvement. These practices include parents' engagement, curriculum reform, teachers' well-being, and professional development for both teachers and leadership members. This case study is an evidence-based guideline for educators and decision makers seeking quality education in their personalized-context learning community.

Keywords School improvement · Quality education · Leadership practices

1 Introduction

1.1 Statement of the Problem

In recent years, deliberate educational policies and strategies have been implemented as attempts to improve students' achievement and reform the whole schooling process in many countries (Fullan, 2009). School leadership practices are considered the most significant factor of the major impacts on students' learning (Hattie, 2009). Moreover, Hargreaves and Shirley (2009) suggest that school leadership has to seek for ceaseless changes to cope with the developing needs of the learners. Since school leaders are responsible for schools' improvement, they need to develop their staff to reach their optimum performance (Ellett & Teddle, 2003) as their teaching approaches and

A. E. Abouelanein (✉) · M. Hossni
Faculty of Education, The British University in Dubai, Dubai, United Arab Emirates
e-mail: 21000466@student.buid.ac.ae

© The Author(s) 2023
K. Al Marri et al. (eds.), *BUiD Doctoral Research Conference 2022*,
Lecture Notes in Civil Engineering 320,
https://doi.org/10.1007/978-3-031-27462-6_5

practices are closely linked with students' achievement and thus the overall school improvement (Lambert, 2003).

1.2 Rationale for the Study

Since The Knowledge and Human Development Authority (KHDA) in Dubai established the Dubai School Inspection Bureau (DSIB) to provide information on the standard of private schools in the emirate of Dubai, schools ratings (weak, acceptable, good, very good, outstanding) have been the major indicator of the quality of teaching and learning each school has (KHDA, 2009). Therefore, school owners, parents, educators, and other stakeholders are becoming close observers of the inspection process and reports. In addition, the researchers have been working in the educational field as teachers and academic heads of departments for more than 15 years and have recently witnessed the improvement of school rating from "acceptable" to "good" soon after a new principal was appointed which is a worthy case study to be examined.

1.3 Aim, Objectives, and Research Questions

The aim of this paper is to investigate the successful practices of school leaders that lead to school improvement. To effectively achieve this, the broad aim of the study is divided into three objectives: to identify school leaders and teachers' perception of school improvement, to look into the different dimensions of school improvement, and to identify the most significant practices of school leadership on school improvement. These objectives are structured as research questions as follows:

1. What are school leaders and teachers' perceptions of school improvement?
2. What are the different dimensions of school improvement?
3. What are the successful practices of the school leadership that lead to school improvement?

2 Literature Review

2.1 Conceptual Analysis

School leadership refers to both managerial and administrative decisions and behaviors by the school governing body to influence students' achievement considering both their needs and desires (Sergiovanni, 2009). Moreover, Cuban (1988) states that there is an obvious distinction between leadership and management as leadership is linked with change while management is regarded as maintaining activity.

School improvement refers to the planned educational change that increases learners' outcomes (Gordon, 2016). It can be also conceptualized as the continuous progress of the school in achieving the educational goals it was established for. Teaching and learning, school environment, equal learning opportunities, clear and focused mission, school-home relationship, and monitoring students' progress are all aspects that can clearly identify school improvement (Bush, 2007).

2.2 Theoretical Framework

The capital theory of school effectiveness and improvement has four major concepts that are closely linked with learners' achievement. According to Hargreaves (2001), these four concepts are: outcomes, leverage, intellectual capital, and social capital. For him, outcomes refer to the achievements of the broad educational goals that are either cognitive or moral. He also defines leverage as the relationship between teachers' input and the educational output. He elaborates on this relationship and links between school improvement and the balance between teachers' efforts and the change in students' intellectual and moral dimensions. He concludes that both intellectual and social capitals are key factors in the school improvement process.

In transformational leadership theory, subordinates are stimulated and inspired to fulfill goals set by their leader (Odumeru & Ifeanyi, 2013). This theory identifies four major components of the transformational leader style: charisma or idealized influence, inspirational motivation, intellectual stimulation, and personal and individual attention (Jung & Sosik, 2002). Charisma refers to the admirable manners according to which the leader behaves and deals with subordinates in different situations. Inspirational motivation is the extent to which the leader can push the followers to reach their optimum performance by setting clear framework and specific objectives. Personal and individual attention identifies the leader's individualized attention to each follower's needs.

2.3 Dimensions of School Improvement

Academic performance, learning environment, and efficiency are the main dimensions of school improvement according to Eastern Kentucky University (2008). These dimensions are examined as follows:

Academic Performance

Curriculum

Curriculum improvement includes leveling up grade level expectations, abolishing ongoing assessment, and replacing end-of-subject exams with periodic summative exams (Winter, 2014). Furthermore, Apple and Jungck (2014) advocate teachers' engagement in the processes of curriculum design and development. They believe that practitioners' inputs in the curriculum are the right approach as multiple perspectives are considered which allow a simple, yet effective curriculum design.

Classroom Instructions and Evaluation
Many education policy-makers worldwide believe that effective classroom evaluation leads to a successful educational process (Barzanò & Grimaldi, 2013). Furthermore, literature has confirmed the effect of school leadership on nurturing teachers' learning and development as a key role of school leadership (Flores, 2004). The purpose of teachers' evaluation is to judge both accountability and improvement. However, tensions and ambiguity might exist by the emerging of the two functions (Flores, 2018). Inconsistent or unfair application of these two functions is regarded as lack of integrity (Campbell & Derrington, 2017). Consequently, a general and specific approach is required when chasing effective teachers' evaluation (Flores & Derrington, 2017).

Learning Environment
School Culture
The impact a principal has on the school is of indirect nature through its culture which in return has a direct influence on students' achievement (Watson, 2001). Consequently, Fink and Resnick (2001) suggest that it is the role of the school leader to maintain a hospitable and welcoming culture that promotes teaching and learning. Gerrard and Farrell (2013) likewise highlight the importance for principals to apprehend the school culture prior to any intended change as they are in the front line of the educational system, and their perception of the appropriate culture is a key factor determining the success or the failure of enhancing teaching and learning.

Student, Family and Community Support
Positive school-family relationships have a vital impact on students' achievement and progress in all levels of education (Morera et al., 2015). Consequently, Gilroy (2018) names parents' involvement in the young learners' education as the most significant key in any desired progress in the quality of teaching and learning. Moreover, Bush (2007) argues that it is the school leadership role to have more parents engaged in the learning of their children which can be monitored through the number of parents keen on attending school-parents' meetings.

Professional Development
The relationship between professional development and quality education was examined in a recent study by Giraldo (2014). The findings he concluded after analyzing qualitative and quantitative data collected from four different instruments showed a dramatic change in teachers' performance and students' achievement. Hannay et al. (2003) add that professional development and performance appraisal need to be consistent to accurately measure the academic growth.

3   Methodology

3.1   Research Paradigm

To get unbiased and truthful answers to the research questions, a pragmatic paradigm is utilized to allow flexibility. Consequently, singular and multiple realities will be concluded from both qualitative and quantitative methods (Creswell & Clark, 2011). Accordingly, this paradigm supports a relational epistemology, non-singular reality ontology, a mixed methods methodology, and a value-laden axiology (Kivunja & Kuyini, 2017).

3.2   Research Design

This paper benefits from the mixed method to reach the best understanding of the research topic. In spite of being relatively new, the mixed method ensures the optimum results, candid analysis, and accurate understanding of the investigated topic (Creswell, 2014). Furthermore, objectivity and subjectivity integrate to emphasize inductive and deductive approaches which give a valued opportunity to segregate concepts from practical perceptions (Morgan, 2007).

3.3   Setting and Data Collection Plan

The setting of this research is a private school in Dubai, United Arab Emirates. Three instruments representing the mixed method will be used to collect data. A questionnaire is distributed to teachers and administrative members to collect data related to the first research question. Moreover, an interview is conducted with the school principal and data collected will be analyzed to answer the third research question. The research also investigates DSIB inspection reports for the school of the academic years 2014–2015 and 2018–2019 to answer the second research question. With this document analysis, data selection is required instead of data collection (Bowen, 2009).

3.4   Population and Sampling

The population of this study is 170 teachers and administrative members representing the teaching and nonteaching staff working in the school along with the school principal. For the quantitative method, 100 participants are involved in the questionnaire, 90 teachers and 10 administrative staff. To increase the efficiency of the research, stratified random sampling technique is implemented. This approach fortifies the study and makes its findings more reliable (Creswell, 2014). Furthermore, since it is a case study, a purposeful sample technique is applied and the school principal is interviewed when conducting the qualitative method.

3.5 Instruments

The first instrument is teachers' and leadership members' perceptions of school improvement questionnaire. The second instrument is a semi-structured open-ended question interview with the school principal. The third instrument is the DSIB school reports of the academic years 2014–2015 and 2018–2019.

3.6 Validity and Sensitivity

Validity is an assessment of the reliability of the quantitative instrument that improves the precision of the data collected and the examination of the findings (Messick, 1995). In this study, the validity of the quantitative approach is obtained as it is taken from a dissertation research paper in the British University in Dubai (BUiD) and published on its website (BUiD, 2016). Similarly, questions of the interview are approved by academics and peer researchers in the same university.

Sensitivity is defined as whether or not the instrument is capable of accurately measuring variability in responses (Zikmund, 2003). To ensure that the study quantitative instrument is sensitive, participants choose from a five Likert-Scale ranging from strongly disagree, disagree, neither, agree, and strongly agree.

4 Results, Analysis, and Discussions

4.1 Data Analysis

For the quantitative method, questionnaires are distributed and filled in as hard copies, and all data is manually uploaded to Statistical Package for the Social Sciences (SPSS) software. The SPSS is utilized to compare between different questions of the questionnaire. For the qualitative method, tabulating technique is used for both the semi-structured open-ended questions interview and the DSIB inspection reports.

4.2   Discussion of Quantitative Data

Responses to the questionnaire questions vary according to each domain. The first domain related to leadership and management shows dissatisfaction among participants about the relationship the school leadership has with them (questions 3–6). However, on the academic level, the overall responses express their agreement with the leadership practices (questions 1, 2, 7, 8).

The second domain of the questionnaire is about teaching and learning approaches of the school leadership. In this section, participants agree with the approaches the school leadership guides them to. On the other hand, the last domain which is about the school culture, responses largely vary from the extreme agreement about the learning opportunities offered to learners to the disagreement about parents' engagement.

4.3   Discussion of Qualitative Data

The DSIB reports of the school inspection show the main areas of improvement of the school in the academic year 2014–2015 and 2018–2019 as follows:

1. Students' achievement improved from 18 "Acceptable" indicators in 2014–2015 to only 3 "acceptable" indicators in 2018–2019 which means that 15 different areas of improvement changed to "Good" in students' achievement.
2. Students' personal and social development enhanced from 8 "Good" indicators in 2014–2015 to "Very good" in 2018–2019.
3. Teaching and Assessment enhanced as 4 "Acceptable" indicators in 2014–2015 changed to only 2 "Acceptable" indicators in 2018–2019 which means that 2 of the areas improved to "Good".
4. Curriculum improvement is significant as it changed from 4 "Acceptable" indicators and 4 "Good" indicators in 2014–2015 to 4 "Good" indicators and 4 "Very good" indicators consecutively in 2018–2019.
5. School leadership and management improved from 4 "Acceptable" indicators and 1 "Good" indicator in 2014–2015 to 4 "Good" indicators and 1 "Very good" indicator consecutively in 2018–2019.

The interview with the school principal reveals several aspects of his leadership practices that are effectively interpreted when linked with DSIB reports. The principal says that the improvement of the curriculum is due to the cumulative efforts of all teachers and coordinators along with the head of academics in the school. He also explains that parents' involvement in the teaching and learning of their children is fostered by the regular teachers-parents and leadership-parents' forms held along with the parents' council that links between parents and the school administration.

The principal also reveals how the school culture is promoted by ensuring a positive relationship between the different subordinates and the school leadership. He also adds that the teachers' council is a crucial step to ensure that teachers' needs and demands are delivered to the governing board of the school. His personal engagement in professional development and class observation is another key behind building a healthy rapport with the teachers along with the regular meetings with the teachers.

4.4 Key Findings

There are some key findings that can be summarized as follows:

1. School principal practices have a significant impact on the overall school improvement.
2. Parents' engagement in their children's learning is a cornerstone in school improvement.
3. Curriculum reform, as a main indicator of school improvement, starts with the practitioners' engagement in its design and development.
4. Data-driven professional development for both teachers and leadership members is vital for school improvement in areas of attention.

5 Conclusion

5.1 Implications of the Current Study

Many studies on the best practices of leadership practices for school improvement have been conducted. To situate this study amongst similar studies, various related works were investigated and their findings are compared with the findings of this research. What makes this study distinct from other studies is that it relies on three different instruments and the cumulative analysis of the collected data is utilized to conclude with the best practices of leadership on school improvement.

5.2 Limitation of the Current Study

Time limit is the major limitation of this case study as it was conducted in a short time. Moreover, teachers' consents were not easily obtained particularly when they knew that they were to evaluate the practices of the school leadership regardless of the researchers' confirmation of the absolute confidentiality.

5.3 Recommendations for Future Research

It is recommended not to limit the qualitative instrument to the school principal as it would be more beneficial to listen to teachers' insights and suggestions for the future research. It is also recommended that a comparative analysis of similar studies in the same context be conducted to best conclude with generalized successful practices of school leadership for school improvement.

References

Apple, M., & Jungck, S. (2014). *Official Knowledge: Democratic Education in a Conservative Age.* London: Routledge.

Barzanò, G., & Grimaldi, E. (2013). Discourses of merit: the hot potato of teacher evaluation in Italy. *Journal of Educational Policy, 28*(6), 767–791.

Bowen, G. (2009). Document analysis as a qualitative research method. *Qualitative Research Journal, 9*(2), 27–40.

BUiD, The impact of leadership styles on teachers' professional development: a study of a private school in Dubai (2016). https://bspace.buid.ac.ae/bitstream/1234/925/1/2014201047.pdf. Accessed 27 May 2019

Bush, T. (2007). Educational leadership and management: theory, policy, and practice. *South African Journal of Education, 27*(3), 391–406.

Campbell, J., & Derrington, M. (2017). High-stakes teacher evaluation policy: U.S. principals' perspectives and variations in practice. *Teachers and Teaching Theory and Practice, 24*(3), 1–17.

Creswell, J. (2014). *Research Design: Qualitative, Quantitative and Mixed Methods Approaches.* California: Sage.

Creswell, J., & Clark, V. (2011). *Designing and Conducting Mixed Methods Research* (2nd ed.). California: Sage.

Cuban, L. (1988). *The Managerial Imperative and the Practice of Leadership in Schools.* New York: State University of New York Press.

Eastern Kentucky University, District Level Performance Descriptors for Kentucky's Standards and Indicators for School Improvement (2008). https://kecsac.eku.edu/sites/kecsac.eku.edu/files/files/SISI08.pdf. Accessed 27 May 2019

Ellett, C., & Teddle, C. (2003). Teacher evaluation, teacher effectiveness and school effectiveness: perspectives from the USA. *Journal of Personnel Evaluation in Education, 17*(1), 101–128.

Ezzani, M. (2015). Coherent district reform: a case study of two California school districts. *Cogent Education, 2*(1), 1–20.

Fink, E., & Resnick, L. (2001). Developing principals as instructional leaders. *Phi Delta Kappan, 82*, 598–606.

Flores, M. (2004). The impact of school culture and leadership on new teachers' learning in the workplace. *International Journal of Leadership in Education, 7*(4), 297–318.

Flores, M. (2018). Teacher evaluation in Portugal: persisting challenges and perceived effects. *Teachers and Teaching Theory and Practice, 24*(3), 223–245.

Flores, M., & Derrington, M. (2017). School principals' views of teacher evaluation policy: lessons learned from two empirical studies. *International Journal of Leadership in Education, 20*(4), 416–431.

Fullan, M. (2009). Large-scale reform comes of age. *Journal of Educational Change, 10*(2), 101–113.

Gerrard, J., & Farrell, L. (2013). 'Peopling' curriculum policy production: researching educational governance through institutional ethnography and Bourdieuian field analysis. *Journal of Educational Policy, 28*(1), 1–20.

Gilroy, P. (2018). Preparing pre-service teachers for family-school partnerships. *Journal of Education for Teaching, 44*(3), 251.

Giraldo, F. (2014). The impact of a professional development program on English language teachers' classroom performance. *PROFILE Journal, 16*(1), 63–76.

Gordon, H. (2016). We can't let them fail for one more day: school reform urgency and the politics of reformer-community alliances. *Race Ethnicity and Education, 19*(1), 1–22.

Hannay, L., Seller, W., & Telford, C. (2003). Making the conceptual shift: teacher performance appraisal as professional growth. *Educational Action Research, 11*(1), 121–140.

Hargreaves, A., & Shirley, D. (2009). *The Fourth Way*. California: Corwin.

Hargreaves, D. (2001). A capital theory of school effectiveness and improvement. *British Educational Research Journal, 27*(4), 487–503.

Hattie, J. (2009). *Visible Learning*. Routledge.

Jung, D., & Sosik, J. (2002). Transformational leadership in work groups: the role of empowerment, cohesiveness, and collective-efficacy on perceived group performance. *Small Group Research, 33*, 313–336.

KHDA, Inspection Handbook (2009). https://www.khda.gov.ae/CMS/WebParts/TextEditor/Documents/DSIB-HANDBOOK_EN.pdf. Accessed 27 May 2019

Kivunja, C., & Kuyini, A. (2017). Understanding and applying research paradigms in educational contexts. *International Journal of Higher Education, 6*(5), 26–41.

Lambert, L. (2003). *Leadership Capacity for Lasting School Improvement*. Alexandria: ASCD.

Messick, S. (1995). Validity of psychological assessment. *American Psychologist, 50*, 741–749.

Morera, M., Expósito, E., López-Martín, E., Lizasoain, L., Asencio, E., & Gaviria, J. (2015). Parental involvement on student academic achievement: a meta-analysis. *Educational Research Review, 14*, 33–46.

Morgan, D. (2007). Paradigms lost and pragmatism regained: methodological implications of combining qualitative and quantitative methods. *Journal of Mixed Methods Research, 1*, 48–76.

Odumeru, J., & Ifeanyi, G. (2013). Transformational vs. transactional leadership theories: evidence in literature. *International Review of Management and Business Research, 2*(2), 355–361.

Sergiovanni, T. (2009). *The Principalship. A Reflective Practice Perspective*. Boston: Pearson.

Watson, N. (2001). Promising practices: what does it really take to make a difference? *Education Canada, 40*(4), 4–6.

Winter, C. (2014). Curriculum knowledge, justice, relations: the schools' white paper (2010) in England. *Journal of Philosophy of Education, 48*(2), 276–292.

Zikmund, W. (2003). *Business Research Methods* (7th ed.). Ohio: Thomson South Western.

Educating Learners with SEND in One Private Mainstream School in Dubai: Effectiveness and Challenges of Policy Implementation

Muntaha Badawieh, Abdulai Abukari, and Eman Gaad

Abstract This study investigates the SEND policy implementation in one private school in Dubai. It highlights whether the school's teachers and support staff adopted and effectively implemented the SEND inclusion policy. The investigation process focuses on three areas which are curriculum modifications, Assessment differentiation, and teacher's professional development. A qualitative research study of semi-structured interviews has been conducted to investigate the effectiveness and challenges of SEND policy implementation; for this purpose, interviews were conducted with the school's teachers for data analysis; the researcher has decided to use the qualitative approach because it shows a subjective evaluation of beliefs, behaviour, and attitudes, the total number of participants were two subject teachers and one school principal. This study investigated all the significant aspects, and the overall result was conducted to provide the proper recommendations. The qualitative study findings indicate that the school policy implementation is very effective, and teachers' awareness of the inclusion policy is appropriate. Nevertheless, they face difficulties in accommodating students' needs in the classroom due to the massive number of students in each classroom. Also, it showed that the inclusion support team acts as a backbone in supporting the teachers through the weekly meetings, which help them modify the curriculum and accommodate the assessment for their students. In terms of challenges, the school has few numbers of inclusion support teachers; therefore, it leads to some challenges teachers face when implementing the policy.

Keywords SEND · Inclusion policy · Inclusive education · Inclusive classroom

1 Introduction

The United Arab Emirates (UAE) has always been interested in developing the education system to achieve global educational standards, and it has sorted Education as a primary priority. The UAE government followed several steps to promote inclusive

M. Badawieh (✉) · A. Abukari · E. Gaad
Faculty of Education, The British University in Dubai, Dubai, UAE
e-mail: 20003293@student.buid.ac.ae

© The Author(s) 2023
K. Al Marri et al. (eds.), *BUiD Doctoral Research Conference 2022*,
Lecture Notes in Civil Engineering 320,
https://doi.org/10.1007/978-3-031-27462-6_6

Education within the seven emirates. Also, it has released many laws and legislation to protect the rights of students with special needs and disabilities and guarantee them educational rights in the regular school setting (Anati, 2013). Moreover, the United Arab Emirates inclusion education philosophy aims to protect the educational rights of students with special conditions and guarantee their opportunity to enroll in a regular school setting (Gaad & Almotairi, 2013). In 2006 the UAE initiated the first federal law, number 29, to protect the rights of people under the special educational needs umbrella, including their social and educational rights, and ensure the best services are provided to them (Gaad, 2010). Furthermore, the Ministry of Education (MOE) in Dubai had implemented an initiative called the "School for All" in 2010, which states that all students with special needs and disability have equal rights in education in a mainstream school setting, regardless of any disability and conditions may have (Sheikh, 2015). As a result, the Knowledge and Human Development Authority (KHDA) and MOE implemented the Dubai Inclusive Policy framework in 2017. Which provides all education providers, local supervisors, and policymakers with clear guidelines and standards to empower and support the students of determination (SEND) within a regular classroom setting (KHDA 2017).

Furthermore, inclusion in education encourages fairness and gives equal rights to all children despite any abilities and disabilities; all children are equal and have education rights in the mainstream classroom setting with the proper support (Gaad & Almotairi, 2013). *Inclusion* is defined as a process that considers and responds to the unique needs of all students by increasing communication and participation in learning (UNESCO 2020). Additionally, equity in education stems from the principle of inclusion, which asserts that all students, regardless of their unique needs or impairments, have the right to an education in a typical classroom environment (Gaad & Almotairi, 2013). Moreover, the government of the United Arab Emirates (UAE) recognizes the value and need of integrating children with special educational needs and disabilities (SEND) into mainstream schools (Keis, 2020), designated as People of Determination (POD) in UAE.

1.1 Definitions of Key Terms

First term is Special Educational Need and Disability (SEND): "when a student has special needs related to a specific impairment, which requires the school to intervene to provide the proper support in accessing the school setting and curriculum and reduce any potential related to the disability in a regular school setting with peers with same age "(KHDA 2017, p.8). Second term is Inclusive classroom: is a class where all the students within the classroom accept each other despite their differences. Moreover, it offers an opportunity to put the students in small groups according to their ability to understand that they are doing different classwork (Jeremy, 1999). Thus, inclusive practices must be done by teachers within the regular classroom setting.

1.2 Study Purpose

The study focuses on determining whether the school's teachers and support staff adopted and effectively implemented the SEND inclusion policy or not; for this purpose, interviews were conducted with the school's teachers for data collection. The primary question leading this research is: ***To what extent do the school's teachers and support staff adopt and effectively implemented the SEND inclusion policy?*** However, it is essential to split the major research question into several sub-questions to make the analysis process clear and help to facilitate a systematic approach to the paper; the following are the research sub-questions: (Q1) What does the inclusion policy entails? (Q2) To what extent does the school's teachers and support staff adopt and effectively implemented the inclusion policy? (Q3) What are the recommendations to improve inclusive education in the school? The investigation process focuses on three areas which are curriculum modifications, assessment differentiation, and teacher's professional development.

2 Literature Review

2.1 Inclusion in Education

Inclusion in education requires changing the mindset of society (Acedo et al., 2009). So, education for all means education for everyone without exception (Anati & Ain, 2012). Achieving this is everybody's responsibility to prepare teachers and provide them with the appropriate professional development. School teachers and support staff must be given the tools, training, and equipment to achieve the inclusion policy aims. However, inclusive education means every learner with special needs should feel welcome, safe, accepted, happy, and valued in the mainstream educational setting despite his or her disability. In inclusive education, the children within the same classroom help and interact with each other (Jeremy, 1999), making the image very clear for students with special needs to feel welcome and essential. All the mainstream schools must ensure that all unique and determined learners (SEND) have equal opportunities for high-quality Education to develop students' best potential, they must be included in the regular classroom setting regardless of their differences (KHDA 2017). Moreover, the inclusive policy focuses on minimizing the academic gaps for SEND learners by accommodating their needs in the classroom, which is for all without exceptions (Ainscow, Booth & dydon 2004). The term "equity in education" means everybody has equal educational rights (Gaad & Almotairi, 2013).

2.2 Schools Toward Inclusion

Inclusion is "a process that helps reduce barriers to learning that limit learners' presences, achievement, and participation." Equity is "The education for all learners are equal" which means that every learner matters and matters equally (UNESCO 2020). The Salamanca Conference Statement (2020) concluded that: "Regular schools with inclusive legislation significantly more effective in creating welcome society, and guarantee equal and equitable education for all."(UNESCO 2020). Furthermore, inclusive schools must reduce the barriers to learning and encourage students' progress for all children within the inclusive classroom. Moreover, including all learners means including children with special conditions in the education system, which is an important strand within the international policy agenda, reflecting that inclusion is for all. Also, the United Nations Convention on the Rights of Persons with Disabilities (2008) states: "The inclusive education right includes a convey practices, culture, and policy in all educational settings by adapt or accommodate the various requirements for each student with special needs."(UNESCO 2020).

3 Methodology

The methodology is the term when we find a problem and we want to find the appropriate solution. It indicates how the research was conducted (Taylor, DeVault & Bogdan 2015). This research followed the qualitative approach to investigate the effectiveness of inclusion policy implementation in one private school in Dubai. The qualitative approach allows an in-depth investigation of the topic to improve it (Gaad & Almotairi, 2013). So, this study is focused on how the teachers in the school perceive and implement the inclusion policy. Moreover, the qualitative approach depends on people's behaviours, their actual perspectives, and the experience from their everyday lives (Taylor, DeVault & Bogdan 2015). However, this approach is the proper approach to research as it involves direct interaction between teachers and researchers in a realistic and natural situation. Moreover, it involves the data collected in the same participants setting (Taylor, DeVault & Bogdan 2015). Moreover, the qualitative methodology approach aims to understand and explore other people's thoughts on human or social problems (Creswell, 2017). This chapter consists of four sections: first, the research site and access describe how the researcher collected the data. Second, the research participants. Third, data collection tools. Lastly, the data analysis.

Table 1 Study Participants (TNR 9pt)

Teachers	Gender	Department
P-1	Female	Mathematics
P-2	Female	English
P-3	Female	School Principal

3.1 Research Site and Access

This study was administered in one school that follows the American curriculum, the total number of students is 2,060, and the total number of teachers is 147. The school has its inclusion vision and mission, which directly align and emerge with the Ministry of Education's vision and mission for inclusion. However, the researcher took permission from the school's principals to access the official school inclusion policy document before the data collection. Furthermore, the researcher analyzed the school inclusion policy by focusing on five areas: SEND student identification, Progress monitoring and tracking, curriculum modifications and adaptation, Assessment accommodations, and teachers' professional development sessions.

3.2 Research Participants

The total participants in this study were 3; two subject teachers and one school principal. The school has well-qualified teachers. Most of them have a bachelor's degree in education. The below table represents the number of school staff participated in this study (Table 1).

3.3 Data Collection Tools

The researcher followed the qualitative approach to gather data from the study participants and has used two instruments to collect the data: semi-structured interviews and official documents.

Official Documents
The researcher used three main sources. First, the Ministry of education federal law number 29 of the year 2016. Second is Dubai's inclusive policy framework for school provision (2017). The third is the school's inclusion policy document. Additionally, the researcher used semi-structured interviews accompanied by open-ended questions (Creswell, 2017).

Interviews
The interviews were conducted in a regular school setting, and participants were invited by email to confirm whether they could attend the interview or not. The

researcher has prepared a Schedule for interviews which something essential to collect data faster (McNeill & Chapman 2005). However, all interviews were conducted on Microsoft Teams. During the interview, each participant had a chance to speak freely for 10 minutes, and the following questions were used to guide the participants during the interviews: (Q1) "How do you as an educator **implement** the inclusion policy?" (Q2) "What challenges do you face when implementing the inclusion policy?" (Q3) "What are your recommendations to improve inclusive school education?" (Q4) "How do the professional development sessions help you?"

3.4 Data Analysis

Data analysis is about gathering data and summarizing it to get the results (Creswell, 2017). All the data collected from the interviews and the official documents followed thematic analysis and were organized and analyzed using Microsoft Word.

4   Findings

4.1 Documents Findings

School SEND Policy
The document findings indicated that the school follows the guidelines provided by the MOE in the UAE to implement the inclusion policy. Moreover, all the school teachers collaborate to plan the lessons, and they keep in mind the needs of the SEND student in their classroom. However, they always differentiate the learning objectives to meet the student's needs in the classroom. Also, there is intensive support from the inclusion department in supporting the teachers, and they have weekly meetings to monitor the student's progress. The school believes that every child can learn. Its inclusion policy aims to provide effective intervention programs and services to SEND students that reflect the initial local and international standards to prepare them to become productive members of society (UNESCO 2020). The school mainly uses all the available resources to monitor the SEND students' progress.

Students Identification and Progress Monitoring
The school follows the procedures provided by the MOE to identify the students with special needs, which are done by teachers' collaboration with the school head of inclusion. The finding indicates that the teachers are responsible for and accountable for the students they teach. With the support of the Inclusion department, teachers implement the prescribed learning strategies to help students with SEND meet their learning goals. Teachers are also responsible for monitoring and tracking all students' well-being and achievement progress with SEND in their classes. The inclusion team and teachers meet to review student well-being and achievement progress throughout

the year to ensure that teaching is of high quality and differentiated is being applied appropriately. If monitoring and tracking identify that a student with SEND is not making the expected levels of progress, then the IEP team meets to review and discuss possible interventions.

Curriculum Modifications and Adaptations

Modifications change the curriculum or measurement of learning. These may include a reduced number of assignments or an alternative grading system. Modifications change "what" is learned and the content of the grade- specific curriculum, which may potentially reduce learning expectations. Adaption is a modification of the students learning outcomes. The finding indicates that the school teachers and head of inclusion decided whether the student needs curriculum modification or adaptation and all the details included in the student IEP.

Assessment Accommodations

The finding indicates that the school follow the procedure in modifying the assessments for all students with SEND, and all students receive modified assessments paper that fit their level.

Teachers' Professional Development

According to many teachers' feedback about the professional development sessions, the school provides PD sessions to the teachers to increase their awareness about the inclusion policy. Four interviews were conducted with school teachers, and all answers indicate that the awareness sessions provided by the inclusion department are few. However, in some way, sessions help them to stay updated according to new strategies and methods they can use with their students.

4.2   Interview Findings

In the implementation process, teachers play a vital role in building and developing their students' academic skills within the mainstream classroom. Interviews were conducted with 3 participants. First, one of the school principals asked how she saw inclusive practices in the school, and she said: "We see that we are providing the best support for the SEND students to succeed. Also, we follow MOE and KHDA guidelines in implementing the inclusion policy." That indicates that the school cares about the SEND students and implementation of the inclusive policy is effective.

The second interview was conducted with the math teacher; she has five years of experience teaching in mainstream schools. The interview questions focused on the curriculum modification and adaptation, challenges, and PD sessions provided by the school. She said: "I am always keen to collaborate with the inclusion support team to plan and differentiate the learning objectives for the SEND students, and I keep them updated about the students' progress and challenges that might occur in implementing the plan." Which indicates the intensive support provided by the inclusion department is effective. The third interview was with an English teacher with 14 years of teaching experience. Her interview questions focused on the policy implementation, challenges, recommendations, and PD sessions. She said, "The cooperation between our department and the SEND department has played a tremendous role in helping meet the needs of the students under the SEND umbrella. At the beginning of the year, the SEND department holds an informative meeting with inclusive class teachers. Important information about the students' status and performance and expectations regarding planning and progress monitoring are shared with subject teachers so that necessary differentiation is implemented as soon as classes begin. My SEND students attend my classes and are pulled out during exams with one-on-one support."

5 Discussion and Conclusion

The study investigated the effectiveness of implementing the inclusion (SEND) policy in one private school in Dubai. The researcher used the qualitative approach to collect the data, such as interviews and school inclusion policy documents. Most teachers understand the inclusion policy and how they must implement it, which goes back to the intensive support provided by the inclusion department. The results indicate that the school policy implementation is very effective, and teachers' awareness of the inclusion policy is appropriate. But, they face difficulties in accommodating students' needs in the classroom due to student number in the school as in each classroom, there are 29 students, so the teacher can't give one-on-one support to the SEND student as she will be busy with the whole class. Also, the number of SEND teachers is few and inadequate for the number of SEND students. In curriculum modification, the support team acts as a backbone in supporting the teachers through the weekly meetings, which help them modify the curriculum and accommodate the assessment for their students. In terms of Challenges, the school has few numbers of inclusion support teachers. Moreover, the number of push-in sessions is very few as well. After discussing the results, several recommendations can be proposed for policy implementation to become more effective.

6 Recommendations

At the beginning of the study, the researcher found that the MOE and KHDA provide clear guidelines for the schools to implement Inclusive Education in practice to protect the education right of all SEND students. This study clearly showed that the school follows the guidelines provided by the Ministry of Education and KHDA. Moreover, all the teachers know the importance of including SEND students in the classroom academically, socially, and emotionally. However, due to the insufficient number of support teachers and the growing number of students in the classrooms, teachers cannot support SEND students regularly. Therefore, it would be better for the school to increase the number of SEND support teachers and pull-out sessions.

References

Acedo, C., Ferrer, F., & Pamies, J. (2009). Inclusive education: open debates and the road ahead. *Prospects, 39*(3), 227–238. https://link.springer.com/article/10.1007/s11125-009-9129-7.

Ainscow, M., Booth, T., & Dyson, A. (2004). Understanding and developing inclusive practices in schools: a collaborative action research network. *International Journal of Inclusive Education, 8*(2), 125–139. https://www.tandfonline.com/doi/abs/10.1080/1360311032000158015.

Al Shehhi, K. (2016). Implementing the Inclusive Education Policy in Three RAK Primary Government Schools: An Investigation study. M.Ed. Thesis. The British University of Dubai

Anati, N. M., & Ain, A. (2012). Including students with disabilities in UAE schools: a descriptive study. *International Journal of Special Education, 27*(2), 75–85. https://files.eric.ed.gov/fulltext/EJ982862.pdf.

Anati, N. (2013). The pros and cons of inclusive education from the perceptions of teachers in the United Arab Emirates. *International Journal of Research Studies in Education, 2*(1), 55–66.

Creswell, J.W., & Creswell, J.D. (2017). Research Design: Qualitative, Quantitative, and Mixed Methods Approaches, 5th edn. Sage. https://us.sagepub.com/en-us/nam/research-design/book255675. Accessed 8 June 2021

Gaad, E. (2010). Inclusive education in the Middle East, vol. 41. Routledge, New York. https://www.routledge.com/Inclusive-Education-in-the-Middle-East/Gaad/p/book/9781138866737. Accessed 17 June 2021

Gaad, E., & Almotairi, M. (2013). Inclusion of student with special needs within higher education in UAE: issues and challenges. *Journal of International Education Research, 9*(4), 287–292. https://www.clutejournals.com/index.php/JIER/article/view/8080.

Jeremy, P. (1999). "What is Inclusion?" *The Review: A Journal of Undergraduate Student Research, 1*(2), 15–21. https://fisherpub.sjfc.edu/ur/vol2/iss1/5/. Accessed 4 June 2021

Keis, Y., The Impact of Adopting Positive Education and Wellbeing Culture on Learners with Special Educational Needs and Disabilities (SEND) in a Private and Mainstream School in Dubai: An Exploratory Case Study During Uncertain Times, a Perspective from Dubai. Ph.D. thesis, The British University in Dubai (2020). https://bspace.buid.ac.ae/handle/1234/1689. Accessed 6 Feb 2022

Knowledge and Human Development Authority, Dubai Inclusive Education Policy and Framework (2017). https://www.khda.gov.ae/cms/webparts/texteditor/documents/Education_Policy_En.pdf

Sheikh, S., Equal rights for people with disabilities (2015). https://www.mbrsg.ae/home/publications/op-eds/equal-rights-for-people-with-disabilities.aspx. Accessed 4 June 2021

Taylor, S.J., Bogdan, R., DeVault, M., Introduction to Qualitative Research Methods: A Guidebook and Resource. Wiley (2015)

Towards inclusion in education: status, trends and challenges: the UNESCO Salamanca Statement 25 years on. Inclusive Education Initiative (2020). https://unesdoc.unesco.org/ark:/48223/pf0000374246. Accessed 7 June 2021

The Role of Lexical Cohesion in Improving Twelfth Graders' Essay Writing Quality

Abdelhamid A. Khalil, Emad A. S. Abu-Ayyash, and Sa'Ed M. I. Salhieh

Abstract The current study was conducted to examine the role of lexical cohesion in improving the quality of twelfth graders' essay writing. The study specifically aimed at examining the correlation between lexical cohesive devices (LCD) and the quality of written texts as well as investigating the barriers of employing these devices for twelfth graders. The context was a private American curriculum school in the UAE. The present paper adopted the quantitative correlational and the quantitative survey research approaches. Data were collected using document analysis of 30 twelfth graders' essays and an online survey attempted by 113 English teachers. Data were analysed using correlational statistics, multiple linear regression, and exploratory factor analysis. The results indicated that there was a significantly positive, moderate association between cohesive ties and students' essay scores. The results also demonstrated that there was a significantly linear relation between hyponyms and synonyms and students' writing scores although hyponyms had more effect on the writing score than synonyms. The findings of exploratory factor analysis identified three factors as major barriers encountered by learners while using lexical cohesion in their written texts including (1) lack of resources and instructions, (2) impact of L1 interference and (3) limited lexical awareness.

Keywords Cohesive devices · Lexical cohesion · Writing quality · Essay writing · Twelfth graders

1 Introduction

It is often challenging for L2 learners to produce well-written texts in another language due to the complex requirements it imposes (Ampa & Basri 2019). Essay writing is a productive skill that requires an exceptional mastery of many language structures and ranks at the pinnacle of the language hierarchy. In addition, written essays are comprised of sentences that are tightly linked using linguistic units that

A. A. Khalil (✉) · E. A. S. Abu-Ayyash · S. M. I. Salhieh
Faculty of Education, The British University of Dubai, Dubai, United Arab Emirates
e-mail: mr.abdel7amed5050@gmail.com

© The Author(s) 2023
K. Al Marri et al. (eds.), *BUiD Doctoral Research Conference 2022*,
Lecture Notes in Civil Engineering 320,
https://doi.org/10.1007/978-3-031-27462-6_7

foster cohesiveness (Saleh & Bharati 2022). Although cohesiveness is crucial to comprehending written texts, it is difficult for language learners to produce cohesive texts because they lack the necessary knowledge of cohesive ties (Abu-Ayyash 2021). As a result, the primary purpose of this research is to investigate the role of lexical cohesiveness in enhancing essay writing of grade 12 students.

1.1 The Research Problem

Cohesion in writing is a problematic issue among language learners and employing LCD in written texts is a major challenge for L2 learners as well as teachers (Saleh and Bharati 2022). Additionally, the existing literature about lexical cohesion indicates that it is still challenging for L2 learners to employ lexical cohesion effectively in essays, and the writing production seems not cohesive (Abu-Ayyash, 2021). The researchers of this paper have recognized that twelfth graders encounter some issues regarding using LCD in their essays. Therefore, the present paper is an attempt to fill in this gap in the existing literature through examining the role of lexical cohesion on the quality of students' writing and identifying the barriers of employing these devices properly in written texts.

1.2 Rationale and Significance

Professionally, there is a need to educate teachers and learners on the significance of using LCD in writing. Examining the usage of lexical cohesiveness could therefore assist English teachers to identify ways to help their L2 learners advance in this area. In addition, the recommendations of this study could lead the way for English teachers and curriculum specialists to develop better course materials to address the requirements of language learners, particularly Arab students in the UAE. Research-wise, the study is noteworthy due to the dearth of research on coherent devices in personal narrative essays, as the researchers have been unable to identify a single study on this essay form.

1.3 Research Questions

1-Is there any correlation between the total number of LCD and the quality of writing?
2-Is there any relation between specific types of LCD used and the quality of writing?
3- What are the barriers of using LCD in twelve graders' essay writing?

2 Literature Review

This section describes the various types of LCD based on the model of Halliday and Hasan (1976), and the literature review summarizes the findings of previous studies that investigated the role of cohesive devices in written texts.

2.1 Definition of Key Terms

According to Wang and Zhang (2019), **cohesion** refers to the semantic links that are used to recognize texts from non-texts. Halliday and Hasan (1976) divided cohesion in to grammatical and lexical. This paper addresses only lexical cohesion. **Lexical cohesion** is non-grammatical, and it is constructed based on the use of vocabulary. Lexical cohesion falls into two categories: reiteration (antonyms, repetition, synonyms, hyponyms, and meronyms) and collocation. **Antonyms** are defined by Leech (2016) as words that are opposite in meaning. **Synonyms**, on the other hand, refer to words which carry the same or nearly the same meaning. Halliday and Hasan (1976) defined **repetition** as the restatement of the exact vocabulary word or lexical element. **Hyponymy** refers to the lexical relation between two items in which one is general or superordinate, and the other is more specific or subordinate (Wang & Zhang 2019). Paltridge (2012) states that **collocations** means that two or more lexical elements collocate with one another to provide new meaning.

2.2 Theoretical Underpinning

Halliday and Hasan's (1976) model of cohesion and Halliday's Systemic Functional Linguistics (SFL) are the basic theoretical underpinnings of the current paper.

Halliday and Hasan's Model of Cohesion
The model of cohesion set by Halliday and Hasan (1976) is the major backbone for studies related to cohesion and its analysis in language use (Khalil 2019). Based on this model, cohesion is split into grammatical and lexical. The model splits grammatical cohesion to four basic sub-types including substitution, reference, ellipsis, and conjunctions. However, lexical cohesion is subdivided into five main sub-types including repetition, hyponyms, antonyms/synonyms, collocations, and meronyms (Abu-Ayyash & McKenny 2017). The choice of this model is rooted in the fact that it has paved the way for researchers and linguists to study and analyse cohesive ties that exist in a text (Saputra & Hakim 2020). In addition, the model has been used extensively as the most effective cohesion model in several languages such as German and Turkish.

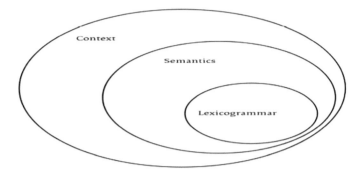

Fig. 1 Stratification in SFL

Systemic Functional Linguistics (SFL)
SFL is a sociocultural theory of language learning devised by Michael Halliday. SFL is composed of five pillars: structure, system, stratification, instantiation, and meta-function (Abu-Ayyash & McKenny 2017). The first three pillars will be briefly introduced since they serve the purpose of the current paper. Firstly, structure is employed to refer to the syntagmatic order of linguistic elements. Moreover, SFL asserts that system addresses the selections made in a language (Martin 2004). The third pillar is stratification through which language is grouped into three coding systems: semantics, lexicogrammar, and sounding or writing as illustrated in Fig. 1 (Halliday & Hasan 1976). According to Martin (2004), SFL has caused radical transformation into language that is now regarded as the source responsible for constructing meaning while grammar is responsible for creating meaning using various wording.

2.3   Related Studies

Correlational Studies Concerning Cohesion Density and the Writing Quality
There is a disagreement among scholars whether the use of LCD correlates with the quality of students' writing (Al-Shamalat & Ghani 2020). For instance, some studies concluded that there was no significant relation between the density of cohesive devices and the quality of students' essays such as (Al-Shamalat & Ghani 2020; Sidabutar, 2021). Conversely, several studies stated that the density of cohesive devices correlated positively with the quality of students' essays including (Chanyoo 2018; Sanchez 2019). By means of elaboration, Liu and Braine (2005) adopted Halliday and Hasan's (1976) cohesion model to examine if there was an association between the number of cohesive ties and the quality of 50 argumentative essays written by Chinese learners of English. The results showed that there was a significant correlation between the density of cohesive ties and the quality of writing. These findings were harmonious with a number of studies (Mora et al. 2021). However, Crossley and McNamara (2010) conducted a study using 184 participants, and they

reached contradictory results in that cohesive devices were not necessarily predictors of the writing quality.

The Challenges of Using Cohesive Ties in Essay Writing

Saleh and Bharati (2022) reported that EFL students encountered challenges related to awareness of LCD and text features. Similarly, Ong (2011) stated that L2 learners encountered some difficulties regarding lexical cohesion including unnecessary addition, omission, misuse and excessive repetition. These challenges were in line with those identified by Saputra and Hakim (2020) in that lack of familiarity of cohesive devices was the most significant barrier of employing these ties effectively in written texts. Likewise, Kirana et al. (2018) concurred that lack of cohesion awareness and improper implementation of language features were two main difficulties faced by EFL students. Consistent with the above findings, Ahmed (2010) and Khalil (2019) concurred that most non-native speakers of English tended to repeat the same lexis in their writing due to insufficient lexical knowledge and the inability to use lexis in producing written texts.

Although an extensive body of research was conducted on the effects of cohesive devices on improving EFL students' written essays, there is scarcity of research carried out on the effects of lexical cohesion on promoting EFL Arab students' essays, particularly personal narrative essays (Sidabutar 2021). The gap in the existing scholarship regarding studies of the relatedness between lexical cohesion and EFL students' writing quality was further acknowledged by recent scholars (Abu-Ayyash 2021; Khalil 2019). Therefore, the current study will bridge the gap in the existing literature by investigating the role of lexical cohesion on improving EFL students' written essays.

3 Methodology

3.1 Research Approach and Design

The present paper adopts the quantitative correlational and the quantitative survey approaches. The quantitative correlational approach serves the purpose of the study since **RQ1** and **RQ2** are focused on investigating the correlation between lexical density and the writing quality (Creswell 2009). However, the quantitative survey approach aligns well with **RQ3** which aims at gathering data about the barriers of using lexical cohesion through an online survey attempted by English teachers.

3.2 Research Instruments

Primary data regarding **RQ1** and **RQ2** were collected through analysing 30 written essays of twelve graders using the linguistic manual annotation model. However, data for **RQ3** were collected using an online survey which consisted of 29 different statements and was developed based on the qualitative findings in the existing literature that identified several challenges encountered by L2 learners regarding use of LCD.

3.3 Sampling

Convenience sampling techniques were used for **RQ1** and **RQ2** since the researchers aimed to analyse lexical cohesion of 30 twelfth graders essay writing. The researchers involved all twelve graders at school and the sample size was (N = 30). However, simple random sampling was selected for **RQ3** in which the researchers used an online survey attempted by (N = 113) English teachers in the UAE. The choice of simple random sampling was beneficial since it was easier to generalize the results as it sought representatives from the wider populations.

3.4 Data Analysis

Data for **RQ1** were analysed using *Spearman rank-order correlation* to examine the degree of association between the total number of LCD and the writing quality as illustrated in Fig. 2.

In addition, multiple regression analysis using SPSS was adopted in line with **RQ2** to investigate the linear relation between each type of LCD and the writing quality. Lastly, exploratory factor analysis (EFA) using SPSS was employed with **RQ3** to identify the main barriers that students encountered while using LCD in writing. The choice of EFA was to examine the correlation among different variables to identify common factors.

4   Results and Discussion

The findings of the study were reported, analysed, and interpreted according to the order of the research questions.

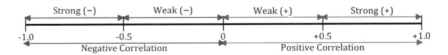

Fig. 2   Basic spectrum of interpreting correlational coefficient

Table 1 Results of the spearman rank-order correlation test

Correlations

			Total number of cohesive devices	Students' mark
Spearman's rho	Total number of cohesive devices	Correlation coefficient	1.000	.401
		Sig. (2-tailed)		.028
		N	30	30
	Students' mark	Correlation coefficient	.401	1.000
	N	Sig. (2-tailed)	.028	
		N	30	30

* Correlation is significant at the 0.05 level (2-tailed)

4.1 Research Question 1

Is there any correlation between the total number of cohesive devices used and the writing quality?

Using the non-parametric test of *Spearman rank-order correlation*, the researchers aimed to find out the correlation between the two variables. Table 1 below presents the results of research question 1.

According to Table 1, it was noted that the *Spearman correlation coefficient r* = 0.401 and the p-value = 0.028. Therefore, the results revealed that there was a moderate positive correlation between the independent variable of the total number of cohesive devices and the dependent variable of students' marks. Besides, the degree of association was statistically significant since the P-value of 0.028 is <0.05. These results were harmonious with several previous studies in that cohesive density led to improving the quality of students' writing (Chanyoo 2018; Liu & Braine 2005; Sanchez 2019). Conversely, these findings contradicted those concluded by other researchers who concluded that there was no statistically significant correlation between cohesion density and the writing quality including (Al-Shamalat & Ghani 2020; Sidabutar 2021).

4.2 Research Question 2

Is there any relation between specific types of LCD used and the writing quality?

In order to answer **RQ2**, the researchers employed multiple linear regression because it works effectively to analyse the relation between a single dependent variable (students' marks) and multiple independent variables (repetition, synonyms, antonyms, hyponyms, and collocation) as concurred by Williams et al. (2013).

Table 2 Coefficients' results of the multiple linear regression test

Coefficients[a]

Model		Unstandardized coefficients		Standardized coefficients	t	Sig.	Collinearity statistics	
		B	Std. Error	Beta			Tolerance	VIF
1	(Constant)	7.438	.178		41.715	.000		
	Hyponyms	.310	.065	.668	4.755	.000	1.000	1.000
2	(Constant)	6.908	.265		26.114	.000		
	Hyponyms	.225	.068	.485	3.285	.003	.760	1.316
	Synonyms	.059	.023	.375	2.543	.017	.760	1.316

a. Dependent Variable: Student Mark

Furthermore, upon testing the required assumptions for running the test, multiple linear regression analysis was used along with "the stepwise method" to answer **RQ2**.

As shown in Table 2, it was observed that the independent variables of hyponyms and synonyms were both significant in terms of the dependent variable of students' marks since their p-value was 0.003 and 0.017 respectively. Since both figures were less than 0.05, they were statistically significant in relation to the dependent variable of students' mark. Furthermore, the *Beta Standardized Coefficients* value was 0.485 for hyponyms and 0.375 for synonyms. Therefore, the model showed that with every increase of one standard deviation in the usage of hyponyms devices, the quality of students' writing rose by 0.485 standard deviations. Similarly, with every increase of one standard deviation in the use of synonyms devices, the quality of students' writing increased by 0.375 standard deviations. In addition, it was concluded that hyponyms had more effect on the quality of writing than synonyms. The findings of RQ2 were contradictory to those reached by Wang and Zhang (2019) who stated that repetition was the most common LCD in students' essays whereas hyponyms was the least commonly used. In addition, their model of analysis showed that there was a statistically significant relation between lexical cohesion and students' scores in writing. This is harmonious with the conclusions reached by the researchers of the present paper. In addition, the low rate of hyponymy used in students' essays was mirrored in the findings of some previous studies (Ampa & Basri 2019).

4.3 Research Question 3

What are the barriers of using LCD in grade 12 students' writing?

In order to answer **RQ3**, the researchers employed Exploratory EFA using several stages so that effective conclusions could be drawn. Thereafter, the Principal Component Analysis (PCA) method of factor extraction was run to find out the minimum number of factors that would account for the maximum variance in data using total variation element (see Table 3). EFA resulted in identifying 3 factors that represented the barriers of using lexical cohesion in students' writing. The first factor was *"Lack of resources and instructions"*, the second factor was *"The impact of L1 interference"* and the third factor was *"limited lexical awareness."* More significantly, the findings of **RQ3** were consistent with the existing literature regarding the challenges that students encountered in terms of using LCD in their writing. For instance, the current paper identified limited lexical awareness as one barrier, which is harmonious with the results concluded by other researchers (e.g., Khalil 2019; Liu & Braine 2005). Similarly, the identified factors of lack of resources and instructions and the impact

Table 3 Rotated component matrix of EFA

Rotated component matrix[a]

	Component		
	1	2	3
Effective guidance	.794		
Error correction	.760		
Curriculum impact	.755		
Clear instructions	.754		
Student motivation	.741		
Curriculum design	.739		
Support using collocations	.725		
Effective hyponymy use	.668		
Teachers' cohesion awareness	.647		
Text cohesion	.605		
Insufficient practice		.751	
Collocations dominancy		.730	
L1 interference		.719	
Vocabulary size		.697	
Insufficient vocabulary		.686	
Repetition dominancy		.646	
Mother tongue effect		.607	
Grading cohesive texts		.581	
Cohesive devices' structure		.537	
Text redundancy		.498	
Synonyms dfficulties		.471	
Antonyms' awareness			.851
Collocations' awareness			.794
Forming antonyms			.671
Synonyms use			.631

of language interference on acquisition of lexical cohesion were consistent with the findings stated by Chanyoo (2018).

5   Conclusion and Implications

This paper sought to examine the role of lexical cohesion in improving twelve graders' essay writing. The current paper employed the quantitative correlational and the quantitative survey research approaches. Data were analysed using correlational statistics, multiple linear regression, and exploratory factor analysis. The results showed that there was a moderate positive correlation between LCD and the writing quality, which was statistically significant. In addition, the findings of RQ2 revealed that there was a linear relationship between hyponymy and synonymy and the quality of students' writing although hyponyms had more effect on the quality of writing than synonyms. Similarly, the results of exploratory factor analysis demonstrated that there were 3 extracted factors that represented the challenges of using lexical cohesion in students' writing namely lack of resources and instructions, impact of L1 interference, and limited lexical awareness.

Taken together, the findings of the present paper provide some significant implications that will be feasible in case they are effectively considered. One implication is for English teachers to provide clear instructions and guidance on how to use cohesive devices effectively to maintain text cohesion. Additionally, English teachers are recommended to provide constructive feedback and error correction to maximize students' effective use of LCD in their written texts. Another implication is for curriculum specialists to integrate teaching cohesive devices with sufficient practice for twelfth graders so that they can improve the quality of their writing. In addition, the findings of the present paper will pave the way for future studies that examine text cohesion and coherence as an indication of maintaining a better writing quality. Finally, while this paper is anticipated to contribute to the existing literature on cohesion, further research is essentially needed to identify effective measures to overcome the challenges and barriers identified in the current paper as hindrance to students' writing quality.

References

Abu-Ayyash, E. A. S. (2021). Non-coherent cohesive texts: Lucky's soliloquy in Beckett's waiting for Godot. *Engl. Text Constr., 14*(2), 182–202.

Abu-Ayyash, E. A. S., & McKenny, J. (2017). The flesh and the bones of cohesive devices: towards a comprehensive model. *J. World Lang., 4*(2), 94–117.

Ahmed, A. (2010). Students' problems with cohesion and coherence in EFL essay writing in Egypt: different perspectives. *Lit. Inf. Comput. Educ. J.*, 211–221.

Al Shamalat, R. Y. S. & Ghani, C. A. B. A. (2020). The Effect of Using Conjunction as Cohesive Device on the Undergraduates' Quality of Writing in Argumentative Essays of Jordanian EFL Learners.

Ampa, A., & Basri, D. (2019). Lexical and grammatical cohesions in the students' essay writing as the English productive skills. *Journal of Physics: Conference Series, 1339*(1), 012072.

Chanyoo, N. (2018). Cohesive devices and academic writing quality of Thai undergraduate students. *J. Lang. Teach. Res., 9*(5), 994–1001.

Creswell, J. W. (2009). Mapping the field of mixed methods research. *Journal of Mixed Methods Research, 3*(2), 95–108.

Crossley, S., & McNamara, D. (2010). Predicting second language writing proficiency: the roles of cohesion and linguistic sophistication. *Journal of Research in Reading, 35*(2), 115–135.

Halliday, M. A. K. & Hasan, R. (1976). *Cohesion in English*. Londra: Longman.

Saleh, M., & Bharati, D. A. L. (2022). The use of cohesive devices in descriptive text by English training participants at PST. *Engl. Educ. J., 12*(1), 95–102.

Khalil, A. (2019). *An Investigation of the Use of Lexical Cohesive Devices in Academic Writing Essays of Grade 9 Learners at an American School in Sharjah* (Doctoral dissertation, The British University in Dubai (BUiD)).

Kirana, E., Syarif, H. & Anwar, D. (2018). Students' writing ability in descriptive texts and their problems of using appropriate adjective in SMP. In *International Conferences on Educational, Social Sciences and Technology* (pp. 508–513). Fakultas Ilmu Pendidikan UNP.

Leech, G. (2016). *Principles of pragmatics*. Abingdon: Routledge.

Liu, M., & Braine, G. (2005). Cohesive features in argumentative writing produced by Chinese undergraduates. *System, 33*(4), 623–636.

Martin, J. R. (2004). Grammatical structure: what do we mean. *Applying English Grammar: Functional and Corpus Approaches* (pp.57–77). London: Hodder Arnold.

Mora, P. A. F., Coyle, Y. & Becerra, J. A. S. (2021). Cohesion in the narrative writing of young EFL learners: correct and incorrect use of local cohesive ties. *Atlantis. J. Span. Assoc. Anglo-Am. Stud.*, 154–177.

Ong, J. (2011). Investigating the use of cohesive devices by Chinese EFL learners. *Asian EFL J. Q., 11*(3), 42–65.

Paltridge, B. (2012). *Discourse analysis: an introduction*. London: Bloomsbury Publishing.

Sanchez, S. (2019). The Use of Discourse Markers in Argumentative Essays by Learners of Spanish as a Foreign Language. (Master thesis), University of Alberta, Canada

Saputra, A., & Hakim, M. A. R. (2020). The usage of cohesive devices by high-achieving EFL students in writing argumentative essays. *Indones. TESOL J., 2*(1), 42–58.

Sidabutar, U. (2021). An analysis of lexical cohesion on the students' writing. *JETAL: J. Engl. Teach. Appl. Linguist., 2*(2), 62–67.

Wang, J., & Zhang, Y. (2019). Lexical cohesion in research articles. *Linguist. Lit. Stud., 7*(1), 1–12.

Williams, M. N., Grajales, C. A. G., & Kurkiewicz, D. (2013). Assumptions of multiple regression: correcting two misconceptions. *Practical Assessment, Research and Evaluation, 18*(1), 11.

Critical Thinking Skills Profile of High School Students in AP Chemistry Learning

Gilan Raslan

Abstract From classrooms to workplaces, educators and policy makers have emphasized the necessity of graduating students who are strong critical thinkers for nearly 50 years and more (Forawi 2016). Critical thinking skills are a vital pillar skill to tackle the challenges of the twenty-first century.

Critical thinking is defined as a set of fundamental skills that must be mastered before one may progress to more complicated thinking. Aiming to obtain more insight into the aspects of critical thinking, the present study particularly examines quantitively the critical thinking skills level of grade 12 students in a scientific learning context. Over a 35-min test, based on Danczak DOT criteria, data was collected and analyzed. The study's findings revealed that the students' critical thinking abilities are in medium range. However, other implications regarding curriculum modifications, educational teaching strategies and teachers' readiness are needed to foster students' critical thinking skills.

Keywords Critical thinking · DOT test · Scientific learning · Teachers' readiness

1 Introduction

Skills matter, and poor skills severely hinder access to better-paying and more gratifying professions, according to a recent study conducted by the Organization for Economic Co-operation and Development (OECD 2016, 2018). Unsurprisingly, critical thinking skills, or CTS, have become a fundamental educational focus in recent decades (OECD 2016; Forawi 2020; Starichkova, Moskovskaya and Kalinovskaya 2022). Because CTS acts as a catalyst, students are able to go beyond simply gathering knowledge to developing a deep grasp of the information offered to them (Amin and Adiansyah 2018; Setyawan and Mustadi 2020). As a result, its most significant

G. Raslan (✉)
Department of Master in Education, Faculty of Education, British University in Dubai, Dubai, United Arab Emirates
e-mail: 20002994@student.buid.ac.ae

© The Author(s) 2023
K. Al Marri et al. (eds.), *BUiD Doctoral Research Conference 2022*,
Lecture Notes in Civil Engineering 320,
https://doi.org/10.1007/978-3-031-27462-6_8

contribution is to promote good decision-making and problem-solving in real-world settings (Perez 2019; Forawi 2020).

Critical thinking CT is a reflective decision-making process that includes critical analysis based on relevant and accountable evidence and justifications (Hasan et al. 2020). Critical thinking is not the same as just thinking. It's metacognitive, meaning it includes thinking about your own thoughts (Mai 2019).

According to Hidayati and Sinaga (2019), critical thinking necessitates logical and interpretative cohesiveness in order to detect prejudices and incorrect reasoning, and it is essential that students learn it.

Learning in the twenty-first century requires a shift in learning orientation, meaning mastering the content of knowledge, skills, expertise (Miterianifa et al. 2021). Students must also have thinking ability, action, and living skills in order to learn in the twenty-first century. One of the life skills is the ability to think critically, and students must have this ability in the twenty-first century, according to the Partnership for 21st Century Skills (Saleh 2019). In addition, students at the postsecondary level and in the workplace require learning assessment and critical thinking abilities in the 21 st century (Forawi 2020).

The major interest of future-oriented scientific, current, and chemical education is to develop students' potential to think critically in all aspects of life (Sadhu et al. 2019). Critical thinking is also important because it allows students to successfully deal with problems and make a tangible contribution to society. It is one of the most important and well-known skills because it is required of everyone in the workplace of different fields such as leadership, and professions that require making decisions and clinical judgment. As a result, critical thinking is an important talent to be taught and educated (Abazar 2020).

In 1955, College Board established the Advanced Placement (AP) program as a non-profit organization that allows willing and academically qualified students to seek studies in the college-level while still in high school, with the chance of obtaining college credit, advanced placement, or both. Through AP classes in 38 disciplines, students learn to think critically, build good arguments, and understand different sides of a problem, all of which culminate in a hard test. These are abilities that will help them succeed in college and beyond (Conger et al. 2021). The AP Chemistry course gives students a college-level foundation in chemistry that will help them succeed in advanced chemistry courses in the future (College Board 2020; Conger et al. 2021). Students learn about chemistry through inquiry-based inquiries that cover topics including the structure of atoms, interactions and bonding between molecules, chemical reactions, reaction rates and thermodynamics equivalent of a college course (College Board 2020). The AP Chemistry course is meant to be a substitute for the general chemistry course that most students take their freshman year of college. Science practices are essential components of the course framework. These practices are; (1) models and representations, (2) question and method, (3) representing data and phenomena, 4) model analysis, (5) mathematical routines, and (6) argumentation; and they explain what a student should be able to do while discovering course concepts (College Board 2020, p. 13–15). Practices are divided into skills, which serve as the foundation for the AP exam's tasks (College Board 2020).

However, the extent to which those science practice skills help in improving the critical thinking skills of the students, not only to comprehend course and to pass the AP exam, but also for them to spot difficulties, solve those problems, and solve problems in everyday life, is still a question to be answered.

Therefore, the research has a purpose to examine the profile of critical thinking skills of high school students studying AP Chemistry course adopted in an American curriculum school in Dubai, using Danczak-Overton-Thompson Chemistry Critical Thinking Test or DOT test.

The study attempts to answer the following question:

To what extent do the AP Chemistry course foster the development of 12th grade students' critical thinking skills?

2 Theoretical Framework

2.1 Bloom's Taxonomy Theory of Learning

Bloom's Taxonomy and critical thinking go hand in hand (see Fig. 1). Bloom's taxonomy walks students through the process of evaluating material or knowledge critically (Wilson 2016).

Bloom's taxonomy begins with knowledge or memory and progresses through a series of levels of questions and keywords that encourage the learner to act. Education and meta-cognition which is the master level of thinking, require both critical

Fig. 1 Interconnection between Critical Thinking and Bloom's Taxonomy *(Adopted from:* https://bcc-cuny.libguides.com/c.php?g=824903&p=5897590)

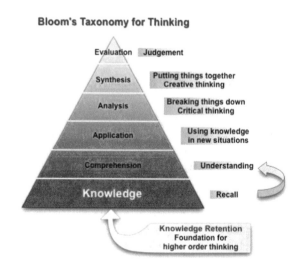

thinking and Bloom's taxonomy (Wilson 2016). Critical thinkers can dissect their own reasoning, draw inferences from available data or apply their understanding of a concept in a variety of ways. They can rephrase questions, divide down tasks into parts, apply information, and generate new data. This is a set of skills that can be taught and learned (Arievitch 2020). Critical thinking, according to Paul, is thinking about one's thinking while he/she is already thinking in order to improve your his/her thinking.

2.2 Critical Thinking and ZPD

Cognitive psychologists were particularly interested in deep thinking and the internal understanding process.

Critical thinking is a cognitive activity that involves the use of the intellect. The ability to transfer knowledge from one discipline to another is referred to as critical thinking. Critical thinking has been linked to the development of individual pondering skills such as logical reasoning and personal judgment, as well as the support of suspicious thoughts (Santos 2017). According to Vygotsky's cognitive development theory, cognitive skills like critical thinking are socially guided and produced (Stetsenko and Selau 2018). The zone of proximal development (ZPD) by Vygotsky, often known as scaffolding, is a concept used in schools to help students learn new skills. The expert gradually withdraws help as the learner achieves competency, until the student is capable of doing the activity on his or her own. This used to be accomplished by offering the student some suggestions and tips to help him solve the problem, while the teacher remained mute until the solver came up with his own hypothesis after properly understanding the problem. Close observation and reason-guide tests would be followed by hypothesis modifications as essential CT phases (Shah and Rashid 2018).

2.3 Guided Inquiry Model

The guided inquiry learning model is a teaching approach that can be used to help students build problem-solving skills through experience (Nisa et al. 2017). This paradigm has been found to be useful in training and guiding students in their grasp of concrete topics as well as their capacity to create higher-order thinking patterns (Seranica et al. 2018). The goal of inquiry-based learning is to educate learners how to research and explain an event. Orientation, formulation of the problem, formulation of hypotheses, data collection, hypothesis testing, and formulation of conclusions, are the guided inquiry learning phases (Putra et al. 2018) which go along with the CT aspects to be assessed in this study (see Table 1) (Hasan and Pri 2020).

Table 1. The Relation between guided inquiry stages and CT skills

No	Stage of Guided Inquiry	Description	HOTS activities	Aspects of HOTS
1	Presenting the problems	At this stage, a phenomenon is presented so that the students' desire to ask questions and formulate problems arises		
2	Arranging hypotheses	Students make hypotheses that are relevant to the problem and decide hypotheses that will be used as apriority research	Analyzing	
3	Investigating and collecting data	Determine and carry out experiment steps and collect data		Critical thinking skill
4	Analyzing the data	Students process data that has been collected and test hypotheses that have been formulated previously	Evaluating	
5	Reflection or conclusion	Students decide, predict, interpret and explain by making conclusions based on the data analysis	Creating	Creative thinking skill

(Adopted from: Hasan, M. and Pri, S. (2020), p. 4)

3 Literature Review

3.1 *Defining Critical Thinking*

'Critical,' 'Criticicism', and 'Critic' are all derivatives of the ancient Greek term 'Kritikos', which means 'able to authorise, perceive, or decide'. In modern English, a 'critic' is someone whose job it is to pass judgment on things like movies, novels, music, and food. It entails expressing an objective and unprejudiced view about anything (Padmanabha 2021).

Philosophy, cognitive psychology, and educational research are the three domains that dominate the debate over the meaning of critical thinking (see Table 2). The philosophy literature focuses on the generation of an argument or opinion (Hitch-cock 2018). The critical thinking process is found to encourage problem solving and deciding what to do, according to the literature in psychology (Sternberg and Halpern 2020). While the majority of education research concentrates on observing

Table 2. Categorization of critical thinking skills as defined by the philosophy, cognitive psychology and education research disciplines

Philosophers[a]	Cognitive psychologists	Education researchers[d]
Interpretation	Apply likelihoods[c]	Recognising underlying assumptions
Analysis	Argument analysis[c]	Identifying central issues
Evaluation		Evaluating evidence or authority
Inference	Statistical reasoning[b]	
Explanation	Verbal reasoning[b]	Drawing warranted conclusions
	Methodological reasoning[b]	
	Conditional reasoning[b]	
	Decision making[c]	
	Problem solving[c]	
	Thinking to test hypotheses[c]	

(Adopted from: Danczak, S. (2018), p. 4)
[a](Facione 1990), [b](Lehman and Nisbett 1990; Nisbett et al. 1987), [c](Halpern 1993), [d](Dressel and Mayhew 1954)

behaviors. Critical thinking, according to these experts, is defined as "purposeful, self-regulatory judgment that results in interpretation, analysis, evaluation, and inference, as well as explanation of the evidential, conceptual, methodological, criteriological, or contextual considerations on which that judgment is based" (Danczak 2018).

3.2 Development of Critical Thinking Skills

Critical thinking skills are developed at a young age, and the effectiveness of educational strategies for enhancing these skills does not vary by grade level (Abrami et al. 2015).

This conclusion is startling from the perspective of Piaget, which considers young children's cognitive processes to be underdeveloped in comparison to those of older people. Thinking is dependent on experience," Piaget says. "Intelligence is the result of an individual's natural potential interacting with their surroundings," he says, adding that small children know more than he can express. The term "development" refers to the general mechanics of action and thought. However, research reveals that there is no specific age at which a child is cognitively equipped to learn more complicated strategies of thinking (Silva 2008), which is in line with both sociocultural and cognitive learning theories. Social connection, according to Vygotsky, is crucial in the cognitive development process (Padmanabha 2021).

3.3 Approaches for Teaching Critical Thinking

Many studies have found that the best teaching effects occur when students' critical thinking skills are explicitly taught and developed over the course of their studies rather than in a single course or semester (Haber 2020). At K-12 education institutions, pedagogical techniques to developing critical thinking range from writing exercises, inquiry-based projects, flipped lectures, and open-ended practical to gamification and work integrated learning WIL (Danczak 2018). Chemical learning necessitates a thorough grasp of concepts, which serves as a basis for grasping later topics (Taber 2019). Students' knowledge is built based on their learning experiences and is linked to their developmental stage as well as the influence of their surroundings. Linking existing understandings with new insights is one strategy to achieve learning success. The constructivist approach is concerned with this process, which focuses on the learners, fostering inventive thinking and allowing them to reach their full potential (Yezierski 2018).

The guided inquiry learning methodology outperforms traditional learning in terms of critical thinking skills, according to many studies (Mulyana et al. 2018 and Seranica et al. 2018). Students will be engaged in learning and will be taught how to tackle environmental problems through guided inquiry. They claim that students' critical thinking abilities develop step by step in inquiry-based learning, including the processes of recognizing and defining issues, generating hypotheses, designing and performing experiments, and formulating conclusions based on the experimental data. Guided inquiry promotes students to develop scientific thinking habits (see Fig. 2) by encouraging them to be more receptive to new ideas in the group and by teaching them critical thinking skills when teachers engage in question-and-answer sessions and guide students in formulating relevant facts. Students consider the entire process rather than simply the final result (Suardana et al. 2019 and Rambe et al. 2020).

Moreover, cooperative Learning is a set of teaching/learning approaches for assisting students in developing critical thinking skills. Students work together to acquire and practice subject matter aspects and achieve common learning objectives. It entails much more than simply grouping students and hoping for the best. These strategies necessitate greater teacher control. Students are asked to discuss a specific topic or participate in brainstorming exercises. Cooperative Learning is a very formal manner of organizing activities in a learning environment that contains specific features aimed at increasing the participants' ability to learn richly and deeply. Examples of these strategies: Think-Pair-Share, Circle-the-Sage, Timed-Pair-Share, Agree-Disagree Line-ups and Rally Coach (Macpherson 2019).

Fig. 2 Scientific thinking habits (Adopted from: Crockett, L. 2018)

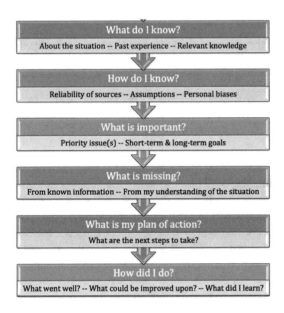

3.4 Importance of Critical Thinking

Is it necessary for us to develop critical thinking skills? What about knowing how to acquire knowledge? In fact, acquiring information is a harmful habit that stands in the way of any discovery. Because, as de Bono puts it, "the illusion of knowledge" will imprison people in what they think they know and prevent them from being open to new ideas (Abazar 2020).

Developing our thoughts is an important element of being educated; it is crucial to a person's development, and every human being has the right to do so. To grow as a well-educated person, our minds must think critically and creatively (Forawi 2020).

Solving complex problems and complicated life issues that necessitate quick and effective solutions is a feature of the 21 st century (Hidayati and Sinaga 2019). The development of students' abilities and competences is in high demand all around the world. Major concerns concerning the capacities of the next generation are regularly acknowledged among educators. Critical thinking, communication, and teamwork abilities are especially important. Schools are obligated to give students with relevant learning opportunities in order for them to develop the skills and competences necessary to succeed in the workplace (Carson 2017).

One of the UAE's main challenges is guaranteeing that its system of education equips students with the skills that the country's developing private market requires, consequently assisting in the diversification of the country's industries and correcting the country's manpower population imbalance. In an innovative economy, the circumstances demonstrate how critical it is for the government to have highly skilled Emirati laborers with significant skill sets available (Forawi 2020). As a result, students' critical thinking skills should be practiced as soon as possible. Junior high school children, with an average age of 11–13 years, are included in the concrete operational cognitive stage, according to Piaget's (1927–1980) cognitive development theory. The idea is that youngsters of that age have been able to use their cognitive skills to identify tangible objects but have not been able to identify abstract objects (Ibda 2015). As a result, kids can begin practicing critical thinking abilities as soon as they enter high school (Hasanah et al. 2020).

3.5 The Assessment of Critical Thinking

According to certain research findings around the world, students' CT skills are still in the poor category (Fadhlullah et al. 2017; OECD 2019; Haber 2020).

The critical thinking assessment is critical because there are various objectives to be met, particularly in science education. Because grasping science information necessitates additional reasoning, CT abilities are required. The importance of critical thinking assessment, according to Ennis, is diagnosing students' CT skills, providing constructive feedback and encouraging students to improve their ability to think critically, as well as inspiring teachers about the suitable teaching strategies needed to teach students CT skills (Hidayati 2019).

The significance of developing students' critical thinking skills at higher education institutions can be seen in its inclusion as a graduate criterion for universities. In addition, research emphasizes the importance of exhibiting critical thinking skills to employers, instructors, and students (Danczak 2018).

The learning outcome can be used to assess the effectiveness of a learning process (Panter and Williford 2017). Critical thinking is difficult to assess. There are features of critical thinking that are both domain-specific and generic (Rashel and Kinya 2021).

The main point of contention in the assessment of CT is whether it is best taught in broad or in specialized disciplines such as history, medicine, law, and education. Critical thinking has been considered as a global, general skill that can be used to any practice of teaching by the 'generalists'. The 'specialists', on the other hand, perceive critical thinking as a skill unique to a certain context and specialty. The discussion over this long-running topic is vital for gaining an insight into the nature of human thought; yet, taking one side or the other is not required. The idea of combining the two approaches has a lot of support. The authors endorse the idea of preparing students for 'multifaceted critical thinking' and the concept of CT that strikes a chord with the pioneers of 'infusion'. (Hidayati and Sinaga 2019).

At universities, critical thinking skills are rarely directly assessed. There are infomercial CT assessments available, which are frequently broad in nature. However, research suggests that evaluations that use a context appropriate to the students' CT skills quite effectively represent their abilities (Chevalier et al. 2020; Wei et al. 2021).

A variety of commercial tools that evaluate critical thinking are available (Assess-mentDay Ltd. 2015; Ennis and Weir 1985; Insight Assessment 2013; The Critical Thinking Co. 2015). The setting of these examinations is generally broad or abstract, and they are created for recruitment purposes. When students, on the other hand, assign meaning to the test environment, a more reliable reflection of students' critical thinking can be derived (Bhutta et al. 2019).

Therefore, for the context of this study, a critical thinking evaluation that tests critical thinking especially from chemistry study is required. According to Suwandi (2011), attainment of advanced thinking skills should not be isolated from assessment, and must be conducted as an integral component of the learning environment to identify students' cognitive growth and learning outcomes, as well as to improve the learning process (Nurfatihah et al. 2021).

4 Methodology

4.1 Design and Methods

This study is quantitative in nature, and aims to examine the critical thinking abilities of class 12 students.

Quantitative research involves the collection of numerical data, and the use of statistics. (Bhandari 2020).

Reflecting on the research question, which focuses on fostering students' critical thinking skills, an assessment tool is used to collect data quantitatively from the students' test results. Then, the test results are analyzed into percentages to measure the causal relation between the quality of the science practice skills implemented in AP Chemistry course and the development of CT skills of high school students.

The paradigm of the study, which is the philosophy that underpins it, is post positivism. Only "fact based" information obtained through using the senses to observe and monitor, including measurement, is considered reliable by this philosophy (Bloomfield and Fisher 2019). In the context of this study, the DOT test results of students are the measurement on which the study's outcomes rely on. In positivism studies, the researcher's role is confined to gather data and analyze it objectively. In other words, while conducting research, the researcher acts as an unbiased analyst who disconnects himself or herself from personal preferences (Bloomfield and Fisher 2019).

4.2 Participants and Ethical Considerations

The participants in this study are 30 twelfth grade students from an American curriculum school in Dubai, adopting American curriculum and AP courses.

Participants were informed that participating in the study was completely voluntary, anonymous, and would have no bearing on their academic records, and that they had the option to withdraw at any moment. All students have been acknowledged with the informed consent. In addition, all techniques were authorized and acknowledged by the school principal.

4.3 Data Collection Instrument

The tool used in this study in a test designed using Google Forms. The test's questions are constructed based on the Danczak-Overton-Thompson Chemistry Critical Thinking Skills Test (Danczak 2018), which is a tool that can be used to assess a student's CT ability at any point during their study of Chemistry. Within a range of quantitative and qualitative reliability and validity testing phases, the DOT test was developed and evaluated throughout three versions. According to the studies, (Li et al. 2020, Salirawati et al. 2021; Susetyo et al. 2021 and Helix et al. 2021) the final version of the DOT test has good internal reliability, strong test–retest reliability, moderate convergent validity, and is independent of past academic success and university of study (Danczak et al. 2016).

The DOT test consists of multiple-choice questions in Chemistry topics to assess five main aspects of CT including: (1) making assumptions: 7 questions (2) analyzing arguments: 7 questions (3) developing hypotheses: 6 questions (4) testing hypotheses: 5 questions (5) drawing conclusions: 5 questions.

A debriefed and revised form of DOT is used in this study, including 15 questions to examine the five critical thinking indicators with three questions for each indicator.

5 Data Analysis and Results

This section depicts the results derived from the DOT examination of student responses.

Data is gathered by including each student's responses to each of the five aspects of the DOT Test.

The students' grades in each of the five key areas are subsequently transformed into percentages (Fig. 3).

The students' critical thinking percentage score is then transformed into qualitative values (categories) based on the following (see Table 3).

The graph (see Fig. 3) below shows the results of students' critical thinking skills exam, which reveal that three components categorized as medium score including 'Developing Hypotheses' (56,6%), 'Testing Hypothesis' (54.4%), and 'Drawing Conclusion' (46.6%), while two components receive scores categorized as low, including 'Making Assumptions' (38.8%) and 'Analyzing Arguments' (36.6%).

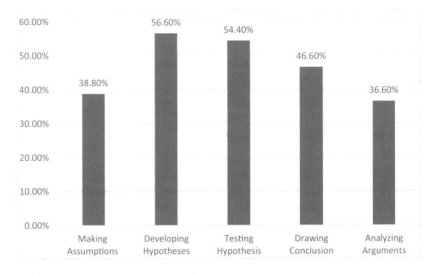

Fig. 3 Percentage of the students' CT skills aspects in DOT-Test

Table 3. Categories of students' critical thinking skills level	No.	Percentage (%)	Category
	1	>80	Very High
	2	>60–80	High
	3	>40–60	Medium
	4	>20–40	Low
	5	<20	Very Low

(Adopted from: Hasanah, S. et al. (2019))

The graph displays the average proportion of students' CT skills from the five components, which is 46.6% which is considered medium. According to the findings, the average outcomes of 12th grade students' critical thinking abilities exams are medium, at 46.6%. This is not in accordance with other studies, which claim that high school students' CT skills are poor (Fadhlullah et al. 2017; Haber 2020).

In the aspects of developing and testing hypothesis of the DOT test, the students demonstrated the ability to predict what will happen in a specific context of interest based on existing evidence and reasoning, then seeking information to confirm or refute this prediction, and lastly drawing a conclusion.

On the other hand, students struggled a bit to postulate and decide the validity of an argument in the aspects of making assumptions and analyzing arguments.

6 Discussion

In discussing the results of the study, three keys with high order abilities were determined to be the greatest in the results of the DOT test: developing hypothesis, testing hypothesis, and drawing conclusion.

Critical thinking skills in the 'Developing Hypotheses' component of students were rated at [49%].

In scientific reasoning, scientists make conclusions based on data, observations, and assumed facts while developing hypotheses. To make a connection or find the intended meaning, an inference is employed to fill in the gaps. These conclusions are not certain, but the hypothesis being constructed has a high level of confidence based on the evidence supplied (Danczak 2018).

The results of the tests suggest that this element is medium, which indicates that students are trained to design a hypothesis through applying the guided inquiry teaching technique as discussed in the literature review.

Results obtained in the section of 'Testing Hypothesis', reflect the same analysis as in the 'Developing Hypothesis' section. With a score of 54.4%, students were able to decide if the idea presented in the passage was supported by the evidence presented, or the deduction had nothing to do with the hypothesis, and there wasn't enough data to back it up.

In a guided inquiry approach, experiments are carried out to test hypotheses (Putra et al. 2018).

Students start with a theory or assertion that they believe is correct and then seek information to corroborate or contradict it. As a result, a premise is formed that is thought to be correct or true (Danczak 2018). This area is very fundamental in science education.

By 46.6% in the area of 'Drawing Conclusion', these results are considered medium; however, it could be considered as 'low' medium. Students may struggle to formulate a conclusion due to a lack of comprehension and inability to make connections. A conclusion's strength is defined by how well the deductions, inferences, and/or premises support it. To reach a conclusion, a scientist will combine multiple pieces of knowledge, such as deductions, inferences, or premises (Danczak 2018). This indicator is consistent with the constructivist approach discussed earlier in the literature review, which emphasizes learners using prior knowledge, encouraging inventive thinking, and allowing them to grow and thrive (Yezierski 2018).

Moreover, formulation of conclusions is one of its essential learning phases in the guided inquiry model (Putra et al. 2018).

The test findings revealed that the area with the lowest score, 36.6%, is 'Analyzing Arguments'.

Students must decide whether or not an argument is valid as part of the scientific process. This necessitates distinguishing assumptions (spoken or implicit), inferences, deductions, and premises, conclusions (certain conclusions in a statement may be implied), and if the argument is relevant to the topic being addressed.

Even if there is sufficient evidence, reliable sources, and supporting material, an argument might be regarded weak if it is unimportant and unrelated to the question being presented (Danczak 2018).

In summary, the average of all components of critical thinking skills is 46.6%, demonstrating a medium category, according to the criteria used. Referring to this research question, this 46.6% average indicates that the science practice implemented in the AP Chemistry course can assist in fostering the CT skills of the high school students.

Whereas, it contradicts the results of the three-year PISA research conducted from 2009 to 2015, which revealed low scores due to students' lack of familiarity with higher-order thinking (Hidayati and Sinaga 2019).

7   Recommendations and Limitations

The exam results are influenced by a number of other factors, such as the process of teaching and learning in the classroom, which is not attuned to developing CT skills in conformance with the expectations of the twenty-first century. Students' inadequate critical thinking abilities are attributable to a lack of activity and training, as well as restricted resources and time, which limit the environment's ability to build critical thinking skills (Fadhlullah et al. 2017).

Memorization should not be prioritized in learning activities (DuDevoir 2018). To solve problems and make judgments, students should be able to derive, interpret, and evaluate information. In the learning process, teamwork and collaboration are also stressed while solving difficulties (Hagemann and Kluge 2017). Learning must also shift from a focus on low-level thinking abilities to one that prioritizes high-level thinking skills (Hasanah et al. 2020).

The study's limitations include the small sample size, making it difficult to generalize the findings and draw firm conclusions based on such a small sample size. To confirm the study results, tt is necessary to conduct a larger sample size study on a broader scope. For example, conduct the study on all grades 10, 11 and 12 students who study Chemistry.

Also, the gender factor can be included in the results and the data analysis. The study conducted by the researcher was on two sets of students, 20 students from the girls' high school section and 10 students from the boys' high school section. Moreover, confounding variables should have been taken into consideration (Jeske and Yao 2020). The environmental conditions of the exam were not identical, since another instructor teaches in the boys section. This instructor may have influenced the students' responses.

Lastly, the study's instrumental tool did not include all the components of the original DOT exam. These metrics may not be able to fully represent all characteristics of an instance.

In summary, the way science curricula are developed will have an impact on future science instruction. This concept is further backed by a significant requirement to incorporate critical thinking skills into science training in order to improve learning outcomes in schools and beyond.

8 Conclusion

The learning experiences that students have, have a big impact on their critical thinking skills. Students will acquire critical thinking abilities if they are frequently offered training to carry out CT activities during the learning process. As a result, future study should emphasize the significance of teaching critical thinking skills to students at an early age, and making it a main priority educational objective. Moreover, teachers should devise teaching techniques that allow students' engagement in activities that assess in the development of critical thinking skills (Chu et al., 2017; Emerson 2019). It is the role of the institutes to keep a closer eye on actual teaching in the classrooms.

Once educated, creative and critical thinking need to be assessed (Abazar 2020). Several instruments are available to help with this, but evaluators must ensure that these instruments are used appropriately in a correct setting, because changes in testing techniques can impact the result's accountability (Forwai 2020). In addition, a study of how science teachers integrate reasoning and critical thinking abilities into teaching and increasing students' learning should be conducted.

Finally, we may firmly admit at the end that critical thinking in science education is the magic wand that will usher in a knowledge-actions based society. That knowledge-actions based society, whether in the United Arab Emirates or elsewhere in the world, will be able to maintain control over the present while deciding on and planning for the future with the adherence to high ethical and moral standards.

References

Amin, A., & Adiansyah, R. (2018). *Lecturers' perception on students' critical thinking skills development and problems faced by students in developing their criticalthinking skills.* Jurnal Pendidikan Biologi Indonesia.

Arievitch, I. (2020). Reprint of: the vision of developmental teaching and learning and Bloom's taxonomy of educational objectives. *Learn. Cult. Soc. Interact., 27*, 100473.

Bhandari, P. (2020). What is quantitative research? | Definition, uses and methods.

Bhutta, S., Chauhan, S., Ali, S., Gul, R., Cassum, S. & Khamis, T. (2019). *Developing a rubric to assess critical thinking in a multidisciplinary context in higher education.*

Bloomfield, J. & Fisher, M. (2019). Quantitative research design. *J. Australas. Rehabil. Nurses Assoc.*

Carson, A. (2017). Problem with problem solving: teaching thinking without teaching knowledge the mathematics educator, vol. 17, no. 2, pp.7–14.

Chevalier, M., Giang, C., Piatti, A. & Mondada, F. (2020). Fostering computational thinking through educational robotics: a model for creative computational problem solving. *Int. J. STEM Educ., 39.* https://doi.org/10.1186/s40594-020-00238-z

Chu, S., Reynold, R., Tavares, N. & Notari, M. (2017). *21st century skills development through inquiry-based learning; from theory to practice.* Singapore: Springer. https://doi.org/10.1007/978-981-10-2481-8

College Board. (2020). *AP Chemistry course and exam description.*

Conger, D., Long, M. & McGhee, R. (2021). Advanced placement and initial college enrollment: evidence from an experiment. *Educ. Financ. Policy* . https://doi.org/10.1162/edfp_a_00358

Danczak, S. (2018). *Development and validation of an instrument to measure undergraduate chemistry students' critical thinking skills.* PhD. Thesis. Monash University.

Danczak, S., Thompson, C. & Overton, T. (2016). Development of a chemistry critical thinking test: initial reliability and validity studies.

Danczak, S., Thompson, C., & Overton, T. (2017). What does the term critical thinking mean to you? A qualitative analysis of chemistry undergraduate, teaching staff and employers' views of critical thinking. *Chem. Educ. Res. Pract., 18*, 420–434.

Dressel, P. L., & Mayhew, L. B. (1954). *General education: Explorations in evaluation.* Washington, D.C.: American Council on Eduction.

Dudevoir, C. (2018). It's not a memory test; education needs to focus on critical thinking.

Emerson, K. (2019). A model for teaching critical thinking. http://files.eric.ed.gov/fulltext/ED540588.pdf

Facione, P.A. (1990). Critical thinking: A statement of expert consensus for purposes of educational assessment and instruction. Executive summary. "The delphi report".Millbrae, CA: T. C. A. Press

Forawi, S. (2020). *Science and mathematics education in multicultural contexts: new directions in teaching and learning.* Champaign, IL:Common Ground ResearchNetworks

Haber, J. (2020). *It's time to get serious about teaching critical thinking.* The MIT Press Essential Knowledge series

Hagemann, V. & Kluge, A. (2017). Complex problem solving in teams: the impact of collective orientation on team process demands. *Front. Psychol., 8*, 1730. https://doi.org/10.3389/fpsyg.2017.01730

Halpern, D. F. (1993). Assessing the effectiveness of critical thinking instruction. *J. Gen. Educ., 50*(4), 238–254.

Hasan, M. & Pri, S. (2020). Implementation of guided inquiry learning oriented to green chemistry to enhance students' higher-order thinking skills. *J. Phys..* https://www.researchgate.net/publication/339680788_Implementation_of_guided_inquiry_learning_oriented_to_green_chemistry_to_enhance_students'_higher-order_thinking_skills

Helix, M., Coté, L., Stachl, C. & Baranger, A. (2021). Measuring integrated understanding of undergraduate chemistry research experiences: Assessing oral and written research artifacts. *Chem. Educ. Res. Pract.*

Hidayati, Y. & Sinaga, P. (2019). The profile of critical thinking skills students on science learning. *J. Phys.*

Hitchcock, D. (2018). Critical thinking. *The Stanford Encyclopedia of Philosophy* . https://plato.stanford.edu/archives/fall2018/entries/critical-thinking/

Jeske, D. & Yao, W. (2020). Sample size calculations for mixture alternatives in a control group vs. treatment group design. *Stat. J. Theor. Appl. Stat., 54*, 1–17. https://doi.org/10.1080/02331888.2020.1715407

Lehman, D. R., & Nisbett, R. E. (1990). A longitudinal study of the effects of undergraduate training on reasoning. *Developmental Psychology, 26*, 952–960.

Li, Y., Li, X., Zhu, D. & Guo, H. (2020). Cultivation of the students' critical thinking ability in numerical control machining course based on the virtual simulation system teaching method. *IEEE Access, 8*, 173584–173598. https://doi.org/10.1109/ACCESS.2020.3025079.

Li, Z., Cai, X., Kuznetsova, M., & Kurilova, A. (2022). Assessment of scientific thinking and creativity in an electronic educational environment. *Int. Sci. Educ.*, 1–24.

Miterianifa, M. & Ashadi, A. & Saputro, S. (2021). A Conceptual framework for empowering students' critical thinking through problem based learning in chemistry. *J. Phys.* 012046. https://doi.org/10.1088/1742-6596/1842/1/012046.

Mulyana, S., Rusdi, R. & Vivanti, D. (2018). The effect of guided inquiry learning model and scientific performance on student learning outcomes. *Indones. J. Sci. Math. Educ., 2,* 105–109. https://doi.org/10.1080/02331888.2020.1715407

Nisa, E. T., Koestiari, T., Habibbulloh, M. & Jatmiko, B. (2017). Effectiveness of guided inquiry learning model to improve students' critical thinking skills at senior high school. *J. Phys.*

Nisbett, R. E., Fong, G. T., Lehman, D. R., & Cheng, P. W. (1987). Teaching reasoning. *Science (New York, N.Y.), 238,* 625–631.

Nurfatihah, R., Nahadi, N., & Firman, H. (2021). Development of chemistry tests for students on reaction rate subject matter based on critical thinking skills using framework DOT test. *J. Educ. Sci., 5*(4), 702–711.

Padmanabha, C. (2021). Critical thinking: conceptual framework. *J. Educ. Psychol., 11*(4), 45–53, (2018). https://ssrn.com/abstract=3772743

Planter, A. & Williford, L. (2017). *Introduction to Student Learning Outcomes Assessment for Continuing Program Improvement: study guide.* University of North Carolina at Chapel Hill.

Putra, B., Prayitno, B., & Maridi, M. (2018). The effectiveness of guided inquiry and INSTAD towards students critical thinking skills on circulatory system materials. *J. Pendidik. IPA Indones., 7*(4), 476–482.

Rambe, Y., Silalahi, A. & Sudrajat, A. (2020). The effect of guided inquiry learning model and critical thinking skills on learning outcomes. https://doi.org/10.2991/assehr.k.201124.033.

Rashel, U., & Shimuzu, K. (2021). Development and validation of a test to measure the secondary students' critical thinking skills: a focus on environmental education in Bangladesh. *Int. J. Educ. Res. Rev., 6,* 264–274.

Saleh, S. (2019). Critical thinking as a 21st century skill: conceptions, implementation, and challenges in the EFL classroom. *Eur. J. Foreign Lang. Teach., 4*(1), 1–16.

Salirawati, D., Priyambodo, E., Nugraheni, A. & Basuki, R. (2021). Critical thinking ability of high school students in daily life acid-base concept. *J. Pendidik. SAINS (JPS), 9*(1). https://doi.org/10.26714/jps.9.1.2021.1-13.

Santos, F. (2017). The role of critical thinking. *J. Educ. Pract,. 8*(20). http://archive.org/stream/ERIC

Seranica, C., Purwoko, A., & Hakim, A. (2018). Influence of guided inquiry learning model to critical thinking skills. *IOSR J. Res. Method Educ., 8*(1), 28–31.

Seranical, C. & Purwoko, A. (2018). Influence of guided inquiry learning model to critical thinking skills. *J. Res. Method Educ., 8*(2), 28–31. https://doi.org/10.9790/7388-0801022831

Setyawan, D., & Mustadi, A. (2020). Is hidrorium able to improve the students' critical thinking skills? *Jurnal Prima Edukasia., 8,* 20–28.

Seventika, S., Sukestiyarno, Y. & Mariani, S. (2018). Critical thinking analysis based on Facione (2015). *J. Phys. Conf. Ser.* https://doi.org/10.1088/1742-6596/983/1/012067

Shah, T. & Rashid, S. (2018). Applying Vygotsky to adult learning.

Sternberg, R. & Halpern, D. (2020).*Critical thinking in psychology* (2nd edn.). Cambridge: Cambridge University Press.

Sternod. L., & French. B. (2015). The Watson–Glaser™ II critical thinking appraisal. *Journal of Psychoeducational Assessment Pullman,* USA: Washington StateUniversity.

Stetsenko, A., & Selau, B. (2018). Presentation–Vygotsky's approach to disability in the context of contemporary debates and challenges: charting the next steps (special issue–Vygotsky's defectology). *Educação, 41*(3), 315–333.

Suardana, I., Selamet, K., Sudiatmika, A., Putri, S. & Devi, N. (2019). Guided inquiry learning model effectiveness in improving students' creative thinking skills in science learning. *J. Phys.* https://doi.org/10.1088/1742-6596/1317/1/012215

Susetyo, B., Siswaningsih, W. & Fizky, F. (2021). Development of critical thinking test instruments with problem solving context on the salt hydrolysis material. *Indones. J. Sci. Math. Educ.*

Taber, S. (2019). Progressing chemistry education research as a disciplinary field. *Discip. Interdiscip. Sci. Educ. Res.* https://doi.org/10.1186/s43031-019-0011-z

The Organization for Economic Co-operation and Development (OECD). (2019). *PISA 2018 results combined executive summaries Volume I, II & III.* https://www.oecd.org/about/publishing/corrigenda.htm

Wei, X., Lin, L., Meng, N., Tan, W., Kong, S. & Kinshuk. (2021). The effectiveness of partial pair programming on elementary school students' Computational Thinking skills and self-efficacy. *Comput. Educ., 160.* https://doi.org/10.1016/j.compedu.2020.104023

Wilson, L. (2016). Anderson and Krathwohl–Bloom's taxonomy revised. Understanding the new version of Bloom's taxonomy.

Yezierski, K. (2018). Complexity in thinking: constructivism within chemistry learning.

Moving from the Subject-Based Curriculum to the Skills-Based Curriculum in Abu Dhabi Schools: Does It Lead to Reform? A Theoretical Analysis & Case Study Paper in One of Abu Dhabi Private Schools

Yaser Abdulrahman Ibrahim

Abstract This paper is a theoretical research study examining the feasibility of implementation of skills-based curriculum model as it relates to the UAE in comparison to other countries. Through this Paper, the researcher made a conceptual analysis, revisited the relevant literature, and identified the theoretical framework of curriculum reform in the UAE educational system with a view on the intended development of curriculum from the conventional subject-based model into the more futuristic and aspiring skills-based in Abu Dhabi schooling environment. The results showed that educational authorities in the UAE should place their focus on an enhanced model of skills-based curriculum, generally known in the literature as the "attained curriculum" category, as an enhanced model of skills-based curriculum to achieve meaningful learning.

Keywords Skills · Curriculum · Subject-based curriculum · Skills-based curriculum · The attained curriculum · Skills centred learning · Adoption of change

1 Introduction

1.1 Background

"The World Needs a New Curriculum" - with these words chosen to be the title of his book, Prensky (2014) wanted to summarise the whole story by stating that "few today get the best education the world can offer, because most of today's education is for a context that no longer exits" (p.1). In this sense, educational and curriculum stakeholders are called to help students achieve new goals, explore new content, and therefore attain new skills.

Y. A. Ibrahim (✉)
The Faculty of Education, The British University in Dubai (BUiD), Dubai, United Arab Emirates
e-mail: 20180413@student.buid.ac.ae

© The Author(s) 2023
K. Al Marri et al. (eds.), *BUiD Doctoral Research Conference 2022*,
Lecture Notes in Civil Engineering 320,
https://doi.org/10.1007/978-3-031-27462-6_9

Driven by the need to fulfill skill and employability demands, educational authorities in the UAE inaugurated a national project where efforts were concentrated on the re-imagination and reform of the national school curricula. The UAE educational authorities hoped that these new approaches would signal a radical revolutionary shift away from the conventional and classical memorization-based learning towards what they hoped to be a skills-based education whose outcomes would be students well prepared for living and working in the 21st century (Ridge, Kippels and Farah (2017a, b).

1.2   Problem Statement and Questions

From a UAE school's perspective, there has been a tendency to move away from the conventional way of what and how children are taught, into helping them how to think creatively. This notion looks fascinating, but its implementation looks quite challenging. Curriculum should be organised around student-centred world themes and problems. Many educators have now started to realize that focus on the classical *subject-centred* curriculum would not help children become better, even more competent, players of the world. However, while moving to the skills-based education would seem to be a positive move in the right direction, it lacks a very basic and essential element: none of it is "real" (Prensky 2014, p. 60).

Very few studies have been conducted with regards to the subject of skills-based curriculum reform in the UAE educational and schooling contexts in recent years. Most of these studies, such as (Suliman 2000), Farah and Ridge (2009), Ridge, Kippels and Farah (2017a, b), etc. focused their attention on a brief history of curriculum development in the UAE in general with mere references to the need for a more advanced skills-based curriculum reform. In this study, the researcher highlighted this issue in the sense of examining the achievability of intentions made in the UAE to move towards the skills-based curriculum since most problems and activities that the existing curricula try to tackle are just "made up" and not real problems.

1.3   Questions

The questions which the researcher answered through this paper are as follows:

Q1: What is the end target or purpose of moving from the content or subject-centred curriculum into the skills-based?

Q2: What is really needed to achieve the positive move from the content-centred into the skills-centred learning and therefore curriculum?

Q3: More importantly, does this transformation from the subject-based into the skills-based, if feasible, necessarily lead to real curriculum reform?

2 Conceptual Analysis and Literature Review

2.1 Conceptual Analysis – What is Meant by Curriculum?

Defining the word "curriculum", a thorough review of the field would show different and varying interpretations of the term. More than 120 definitions of the word "curriculum" were found just by Portelli (1987). These definitions were cited by Marsh and Willis (2003, p. 7). Ridge, Kippels and Farah (2017a, b) define the term "curriculum" as an umbrella inclusive of all the concepts, underlying educational theories and philosophies, objectives, content, and functions of the "instructional program" in the classroom, as well as written and other materials needed to support the educational system (p. 1). Richards (2012) defines a curriculum in a schooling context as referring to the complete set of knowledge obtained and acquired by students.

In the view of Van den Akker (2003, 2010), curriculum can be seen as a concept which can be dissected into three major components: *intended* curriculum, *implemented* curriculum, and *attained* curriculum. The intended curriculum ideally consists of the guiding documents which, in the UAE context, are developed and sustained to dictate what should be taught in schools. The implemented curriculum describes the activities that actually happen in the classroom. Finally, the attained curriculum is related to the learning process and assessment activities themselves as conceived by students in their classroom. It is concerned with the skills and values students attain as well as what content they absorb and retain.

Essential 21st Century Skills Needed by Today's Students

There has been a recent call by educators and educational experts that students need improved 21st century skills (Ridge, Kippels and Farah (2017a, b). Without these skills, researchers and experts now believe that students will not be effective players in the world system and the global economy. Without such skills, their preparation and qualification for both academic and professional careers will be inadequate.

To the question "what exactly are these 21st century skills?", many current researchers and organisations have attempted to provide answers. In a report published by Hanover Research Center (2019), participant researchers shed light on the subject defining those skills as major computer and technology related 21st century skills. They also include related skills which go beyond the technological expertise, and refer to literacies and proficiencies which will prepare students to face risks, challenges, and opportunities.

2.2 Review of Relevant Literature

Curriculum Reform in the view of the Glamour of a Dead-Living Shark Artwork and Originality of Saber-Tooth Tiger Curriculum:

On the notion of curriculum change and reform, Granville (2014) considered the impossibility of curriculum change in the mind of the educated drawing upon some of the most famous historical curriculum studies: Damien Hirst's Shark artwork (1991) and the classically most famous old fable-based Sabre-Tooth Curriculum theory, recounted in the form of an extended essay book by Benjamin (1939). Granville (2014) drew on a very interesting simile to the famous artwork of the English artist Damien Hirst: 'The Physical Impossibility of Death in the Mind of Someone Living (1991)' (Brooks 2005). The author believes that reform stakeholders must be aware of how significantly challenging curriculum reform is. He then uses the simile of the famous Abner Peddiwell's Saber-Tooth Tiger Curriculum that is summarised by the fact that according to the "rulers of the tribe", the essence of education is timeless and therefore even though the saber-tooth tiger had long time ago gone extinct, the curriculum centered around protecting humans from its dangers is still valid and applicable. However, while it is true there is always a pressing need for fundamental curriculum reform, Granville (2014) adds, there are also certain areas and aspects of the curriculum that are always proof to change.

Conversely, Prensky (2014) believes the world needs a new curriculum. However, in his book he defends the notion that even though skills and standards are crucial to the learning process, we must focus on "accomplishment-based education" (p. 5). Only when move from the subject-based education we can have better players, more effective thinkers, communicators, and doers in the world. In this sense, he concluded that the world needs a new curriculum. The curricular scheme proposed by Prensky (2014) is unique and based on four key pillars: Effective Thinking, Effective Action, Effective Relationships and Effective Accomplishment. Prensky's theory goes beyond the promises defended by "21st century skills."

Curriculum Development Activities of the UAE Ministry of Education
Shortly after the UAE union in 1972, the MoE faced the major challenge of having a variety of different curricula with varying standards, tailored to suit the influx of different people and cultures coming to the country. In 1979, the MoE finally launched its "National Curriculum Project" with the aim of creating a single UAE national curriculum, which was only implemented nationally in 1985 (Ridge 2009).

According to Nanney (2004), "Student-centered learning environments have a heightened advantage over the traditional teacher- centered, subject-centered environment in that they provide complimentary activities, interactive in nature, enabling individuals to […] move forward into increasingly complex levels of content to further their understanding and appreciate subject matter" (p.1). In this sense, changes in textbooks are driven by what is known as the *intended curriculum*. These changes would lead to changes in the way teachers deliver the curriculum in what is known as the *implemented curriculum*. This is what the researcher defends in this study as being the *attained curriculum*. In a sense, the UAE current intended, implemented, and attained curriculum needs to be harmonized and brought into play together to enable policy makers and curriculum developers succeed in the achievement of skills-based curriculum.

Moving to the Skills-Based Curriculum in Abu Dhabi Schools

Based on the conceptual analysis the researcher made of the word "curriculum" and its relevance to the UAE context, the researcher adopts the "attained curriculum" category introduced by Van den Akker (2003, 2010).

In their Brief, Farah and Ridge (2009) states that the UAE government attempted at the time to adopt two new curriculum approaches on a tentative basis: the all new *standards-based* curriculum in Abu Dhabi schools and the new English-medium curriculum in selected government schools, or what is known as the Madares Al Ghad, or Schools of Tomorrow, throughout the UAE, launched in 2007 for the purpose of providing an up-to-date curriculum with English as the medium of instruction (Warner and Burton 2016). The main aim of UAE authorities at that time was to move away from the classical and archaic constraints of memorization-based school teaching methodology towards a more solid and advanced *skills-based* education which will produce students fully prepared to live in, work and cope with the 21st century (Ridge, Kippels and Farah 2017a, b).

2.3   Theoretical Underpinnings

The study of curriculum change, and reform, can be based on theories such as the curriculum evaluation theory defended by Clements (2007). It suits this study since it deals with the analysis of goals or objectives of the curriculum. This theory would drive into the direction of understanding individual interpretations about curriculum and learning experiences. Similarly, according to Clarke (1987), curriculum theories would fall under one of three major categories: Classical Humanism, Reconstructionism and Progressivism.

With the Classical Humanism and Reconstructionism being outdated and focusing on the standard old-fashioned memorization and subject centred learning processes, the researcher would adopt the Progressivist theory in this study since it "promotes a learner-centered approach to education which attempts to promote the learner's development as an individual with intellectual and emotional needs and as a social being." (Tzotzou 2013, p. 3).

3   Research Methodology

3.1   Research Approach and Design

Creswell (2013 & 2014) recommends that research follows one of the five traditions of qualitative research: narrative, phenomenology, ethnography, case study and grounded theory. In this study, the researcher opted for a mixture of the *narrative* and *case study* qualitative research design where the researcher investigated the theoretical and conceptual challenges of developing a skills-based curriculum in the UAE by re-telling the story of other studies: "narrative research employs re-storying the participants' stories using structural devices, such as plot, setting, activities, climax, and denouement" (Creswell 2014, p. 196). This was the researcher's major instrument of data collection and analysis (Merriam 2002).

The researcher chose to conduct a small-scale quantitatively based case study to support some of his findings throughout the study. This small-scale case study, involving two questionnaire surveys, was conducted with educational expert informants: one principal and two teachers from a private Abu Dhabi international school to seek their answers, interpretations, awareness, knowledge, and conceptions of the subject of this study and its major questions.

3.2   Data Collection Method

As mentioned earlier in this paper, aside from the data and findings the researcher drew from the detailed theoretical and conceptual analyses of this study, the researcher used a small-scale case study where the researcher initially planned to collect supporting data through semi-structured interviews. However, due to the limitations of time and resources, the researcher changed the data collection method into a survey of closed-ended questionnaires.

Creswell (2014) agrees with Merriam (1988) and Marshall and Rossman (1989) "that data collection and data analysis must be a simultaneous process in qualitative research" (p. 209). While doing analyses of the findings of the theoretical and conceptual frameworks of curriculum reform, the researcher used the major questions of the study to draft questions directed to a principal and two teachers, one of mathematics and one of English, at a private international Abu Dhabi school. Their answers were used for simultaneous crosschecking and as supporting evidence of conclusions inferred through the theoretical analysis.

3.3   Instrument Design

The small-scale case study was designed through the distribution of closed-ended questionnaires on a private school principal and two teachers, one of mathematics and one of English. The researcher used the Survey Monkey software tool to send the questions to one principal and two teachers, one of English and one of mathematics. The answers and data resulting from the surveys were then used for analysis against the major questions of the study.

4 Data Analysis, Results and Discussion

Focusing on the theoretical and conceptual analyses of curriculum development in the UAE, the results show that the UAE educational authorities have indeed managed to embrace the importance for the need of curriculum reform. However, speaking of the three major categories of curriculum identified in this study; the intended, the implemented, and the attained; studies show that in the UAE there still has been no synergy made among them.

The answers of the survey participants came to sustain this conclusion since the principal and teachers answered positively to the questions on their awareness of the need for creating a coherence between the three major categories of curriculum to achieve a real move towards the skills-based curriculum. The answers of the principal and teachers also show that to ensure a successful implementation of a new curriculum, there needs to be better training for both teachers as well as school administrators to understand what it means in practice to deliver a skills-based syllabus as opposed to the subject-based one. The answers also show that there must be a shift in teachers' attitude regarding the way they deliver the objectives and standards set for their curricula from the beginning of the year. The answers of participants also showed that teachers need solid training to move away from a textbook-centered curriculum to a student-centered curriculum. Such training needs to sustain the fundamentals of teaching yet widen the scope of teaching process to include the practical and professional (real-world) skills beyond the typical pedagogical topics. As stated, more intensive training and familiarisation processes are required in this case, things that the current curricula system does not hold.

Ridge, Kippels and Farah (2017a, b) managed to highlight this issue of the absence of synergy among the three different categories of curriculum in the UAE. This issue would have its big disadvantages in the teaching and learning process where teachers refrain from the adoption of new student-centered approaches to teaching and would hinder the development and implementation of skills-based curricula. In a sense, the UAE current intended, implemented and attained curricula need to be harmonized and brought into play together so that policy makers and curriculum developers can succeed in their mission towards the achievement of skills-based curriculum.

5 Conclusion and Recommendations

5.1 Conclusion – Summary of Results

From the theoretical and conceptual analyses, the researcher made in this study, it was clearly observed that the adoption of a new revamped skills-based curriculum is a rigorous tool towards a solid and comprehensive educational reform within the UAE educational system. Based on the conceptual analysis the researcher made of the word "curriculum" and its relevance to the UAE context, the researcher conclude that educational authorities in the UAE should really place their focus on the "attained curriculum" category introduced by Van den Akker (2003, 2010) being the type of curriculum that helps all elements of the teaching and learning process to effectively implement a solid and enhanced model of skills-based curriculum. Since the attained curriculum is related to what students have learned, adding the factor of "skills" in this category is what really matters.

Curriculum reform and moving towards the skills-based seems not to be an easy task. According to Prensky (2014), "it involves not just "adding technology" or "adding some 21st century skills, but rather, I believe, that we learn to teach our young people Effective Thinking, Effective Action, Effective Relationships and Effective Accomplishment" (p. 1). The researcher's clearest conclusion of this study is that the key for the attainment of these skills to be developed among learners are the teachers themselves in partnership with stakeholders: modified and enhanced curriculum alone will not do the job. Finally, as stated by Prensky (2014), "21st century skills" curriculum should not be based on made-up concepts, but rather on actual student accomplishments in the real world. What is then really needed is a learning process whose outcomes are of real significance and provide real improvements and solutions for real problems in the world. This should indeed be what a skills-based education, and as such curricula, be all about.

5.2 Recommendations for Further Research

This study was mainly intended to be a theoretical overview of the progress and implications of the implementation of skills-based curriculum in the UAE. The study was initially planned to be conducted through a set of interviews with one school administrator and two teachers. However, a survey was done instead with the answers and findings resulting from them covering only a small sample of informant participants. The researcher's recommendation is that this same study could be piloted with a bigger-scale study that covers more UAE schools and more participants. This further research could be conducted to be mainly a case study covering the findings of this theoretical study.

References

Benjamin, H. R. W. (1939). *Saber-tooth curriculum, including other lectures in the history of Paleolithic education.* New York: McGraw-Hill.

Brooks, R. (2005). *Hirst's shark is sold to America, The Sunday Times,* 16 January 2005. https://www.thetimes.co.uk/article/hirsts-shark-is-sold-to-america-6nb2kssrp7h. Accessed April 17, 2018.

Clarke, J. L. (1987). *Curriculum renewal in school foreign language learning.* Oxford: Oxford University Press

Clements, D. (2007). Curriculum research: toward a framework for research-based curricula. *J. Res. Math. Educ., 38*(1), 35–70.

Creswell, J. W. (2013). *Qualitative inquiry and research design: Choosing among five approaches* (3rd ed.). Thousand Oaks, CA: Sage.

Creswell, J. W. (2014). *Research design: qualitative, quantitative, and mixed methods approaches.* Thousand Oaks, CA: Sage.

Farah, S. & Ridge, N. (2009). *Challenges to curriculum development in the UAE.* Dubai School of Government.

Granville, G. (2014). *The Impossibility of Curriculum Change in the Mind of Someone Educated*: Shark, Sabre-Tooth and Junior Certificate, Chapter 3 in Dolan, R. (2014).

Hanover Research Center. (2019). A Crosswalk of 21 st Century Skills. https://www.hanoverresearch.com/search/?_sf_s=21st%20Century. Accessed May 11, 2019.

Marsh, C. J., & Willis, G. (2003). *Curriculum: alternative approaches, ongoing issues* (3rd ed.). Upper Saddle River, NJ: Pearson Education.

Marshall, C. & Rossman, G. (1989). Designing qualitative research. *Issues Appl. Linguist., 1*(2).

Merriam, S. B. A. (2002). *Qualitative research in practice: examples for discussion and analysis* (1st ed.). San Francisco, CA: Jossey-Bass.

Merriam, S. B. (1988). *Case Study Research in Education: A Qualitative Approach.* San Francis-co: Jossey-Bass.

Nanney, B. (2004). Student centred learning. In *Proceedings of the 3rd European Conference on E-Learning..*

Organisation for Economic Cooperation and Development (OECD). (2016). Education at a glance 2016: OECD indicators.

Portelli, J. P. (1987). Perspectives and imperatives on defining curriculum. *J. Curric. Superv., 2*(4), 354–367.

Prensky, M. (2014). *The world needs a new curriculum.* New York: The Global Future Education Foundation and Institute Ltd.

Richards, J. C. (2012). *Curriculum development in language teaching.* Cambridge: Cambridge Language Education.

Ridge, N., Kippels, S., & El Asad, S. (2017a). *Fact sheet: education in the United Arab Emirates and Ras Al Khaimah.* Sheikh Saud bin Saqr Al Qasimi Foundation for Policy Research.

Ridge, N., Kippels, S. & Farah, S. (2017b). *Curriculum development in the United Arab Emirates.* Sheikh Saud Bin Saqr Al Qasimi Foundation for Policy Research.

Schmidt, W. H., et al. (2001). *Why schools matter: a cross-national comparison of curriculum and learning.* San Francisco, CA: Jossey- Bass.

Suliman, O. M. (2000). *A descriptive study of the educational system of the United Arab Emirates.* (Ed. D. dissertation). University of Southern California.

Tzotzou, M. D. (2013). *A critical analysis and evaluation of the Unified Curriculum for the Foreign Languages (EPS-XG Curriculum).* Aspects Today. Issue No. 36, January-April 2013.

UAE Ministry of Education, College Board, and Cambridge University Press (2017). ASP Elite Stream (flyer).

Van den Akker, J. (2003). *Curriculum perspectives: an introduction.* In Curriculum Landscapes and Trends, (pp. 1–10). Netherland: Springer. https://doi.org/10.1007/978-94-017-1205-7_1

Van den Akker, J. (2010). Building bridges: how research may improve curriculum policies and classroom practices. https://research.utwente.nl/en/publications/building-bridges-how-research-may-improve-curriculum-policies-and.

Warner, R. S. & Burton, G. J. (2016). *A fertile oasis: the current state of education in the UAE.* UAE Public Policy Forum. Mohammed Bin Rashid School of Government.

The Education System in Post-conflict Syria – Examining PIRLS as an International Assessment Measure to Ensure the Quality of Students Achievements

Yaser Abdulrahman Ibrahim

Abstract This paper provides a general contextualisation of assessment and learning criteria and processes within the framework of leadership and management perspective in post-conflict and war affected countries, with Syria as an example. Through this paper the researcher provided an overview of the related academic literature done on the subject as well as evaluating the importance of applying one of the International Assessment Measures (IAMs) as a solution for educational reformation. The paper evaluates the adoption of one assessment measure to support the special and emergency requirements of post-conflict countries such as Syria.

Keywords International assessments · International Assessment Measures (IAMs) · PISA · PIRLS · TIMMS · Post-Conflict Syria

1 Introduction

The paper begins with a general overview of the literature of educational leadership and management issues in conflict areas and war affected countries, with a focus on the latest theories of leadership and management in the field of education. Then, the study defines and provides an overview of the best International Assessment Measures to be used in the field as a solid tool for learning and evaluation in those countries. Finally, the study gives more focus on the ideal mode of assessment to be adopted and tailored in a post-conflict Syria, as a means for the reformation of the hugely broken Syrian educational system, with PIRLS as an option.

Y. A. Ibrahim (✉)
The Faculty of Education, The British University in Dubai (BUiD), Dubai, United Arab Emirates
e-mail: 20180413@student.buid.ac.ae

© The Author(s) 2023
K. Al Marri et al. (eds.), *BUiD Doctoral Research Conference 2022*,
Lecture Notes in Civil Engineering 320,
https://doi.org/10.1007/978-3-031-27462-6_10

107

1.1 Rationale of the Study

This paper may well fit as a part of the broader studies focusing on the subject of education reconstruction in post-conflict communities. Several studies have already been conducted to review the feasibility of education system reformation in post-conflict nations. However, very little studies have dealt with issues of assessment measures and systems followed in emergency situations to reform and achieve a better educational quality.

1.2 Questions of the Study

The general question of this paper will be "Which assessment measure of the three International Assessment Measures would be the most innovative and effective tool in the reformation of the educational system in conflict affected countries? However, the specific questions are "Can PIRLS be utilized in post-conflict Syrian elementary schools as a tool of educational reformation?" and "How can PIRLS be tailored to cope with the special emergency requirements of post-conflict countries?".

2 Literature Review

2.1 Conceptual Analysis

An Overview of International Assessment Measures (IAMs):
Academic research would inform that the teachers are subject to the influences of external assessment practices in their way of teaching (Harlen 2004). However, teachers often use assessments as tools of teaching for their own sake, even if they are not completely tailored to the specific needs or requirements of their teaching process (James 2006). Currently, the most prominent external assessment measures are the international assessment measures of TIMSS, PISA, and PIRLS. The PIRLS mode of assessment, standing for "Progress in International Reading Literacy Study", has been managed by Boston College in collaboration with the Evaluation of Educational Achievement (IEA) every 5 years since 2001. PIRLS targets a sample of fourth grade students. In 2011, 66 countries participated in PIRLS, and in 2016 and 2021 the number was 61 countries.

PIRLS is offered to participating countries every five years to assess students' reading achievement in their fourth year at school. PIRLS 2021 is the fifth and latest assessment in the current trend series, following PIRLS 2001, 2006, 2011 and 2016. The latest PIRLS 2016 and 2021 developed solid tools and instruments to reflect the new changes or international reading assessment trends. Through the recent years, some developing countries have also started to benchmark their performance in the use and application of IAMs against the performance scores of their 'developed' neighbors (Bernbaum and Moore 2016, p. 7).

Impact of Conflict and War on Education & IAMs Usage

For some countries, the outlining lineament in recent years has been one of continuing conflicts and political and social chaos. Additionally, inadequate teachers training, insufficient teaching quality, and the absence of official monitoring systems have all led to the absence of robust assessment and evaluation measures in those educational and schooling systems (Sommers 2002; UNESCO 2000: 15; Bernbaum and Moore 2016). Therefore, there has been an increasing demand for the implementation of solid and subtle assessment measures.

In Syria, given the fact the "[t]he destruction and devastation to human life, property, and infrastructure will take decades to repair" (Jones et al. 2017, p.8), it seems that the whole situation is more complicated. "The challenges facing Syria are immense, including… ineffective governance that particularly affected young Syrians through poor education and joblessness …" (Jones et al. 2017, p. 8). In its Education Report on Syria published in March 2017, the International Rescue Committee (IRC) "has collected data that provide critical insight into reading and math levels of internally displaced and host community children." (IRC Report, 2017, p. 1).

International Assessment Measures usage in Post-Conflict Syria

None of the PIRLS resources and websites shows evidence of Syria taking part in any of the PIRLS assessments. Reports of TIMSS and PIRLS assessments published by IEA show Syria participating only in the TIMSS assessments for math skills and achievements. (Building Knowledge in the UAE Report, p. 13).

"For the data discussed in [this] report, IRC used the Annual State of Education Report (ASER) reading and math tools" (IRC 2017, p 4) as an instrument to test reading and math achievements of Syrian students in certain selected areas. ASER assessment was chosen as the measure in selected areas in Syria because the tools are user-friendly and give the chance to build a solid and factual visualization of early grades' reading and math skills. The 2017 ASER Reading Results by IRC in Syria show that 96% of male and female students are unable to read at Grade 2. However, reading illiteracy results change as we advance in the school level, demonstrating less illiteracy rates among students of Grade 8 of 35% (Fig. 1).

Like the ASER initiative, PIRLS can be utilized as a mode of assessment in post-conflict Syria. In this sense, reports show that the students or pupils in Syrian conflict areas do not need the learning process to focus on reading literacy. In this context, the researcher believes that investigating PIRLS as the ideal mode of assessment can give fruitful results to the quality of their educational system.

2.2   *Theoretical Underpinning*

The subject of international assessment application and implementation in post-conflict context raises some questions about the theories in which assessment, on

Fig. 1 Graph taken from the
IRC Report, 2017

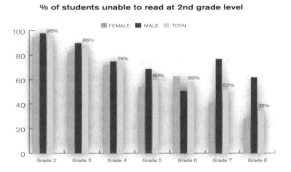

the one hand, and learning and teaching quality, on the other, are expressed (James 2006). This paper of assessment measures would fall within the general theories of behaviourism, constructivism, and socio-culturalism. The reason for this theoretical underpinning classification is the socio-cultural and behavioural relationship between assessment practices and the ways in which students are trained to achieve learning outcomes. It is assumed in academic research that there should be some alignment between classroom assessment and the understanding of social and behavioural contexts of learning (Shi et al. 2016).

2.3 Review of the Relevant Literature

There is a myriad of insights that can be induced from existing literature on the general subject of education in post-conflict societies. There is also a big repository of other scholarly journals, reports and articles that show how post-conflict educational systems have been effectively approached and dealt with. Studies such as Davies (2004a & b), Gallagher (2004), the World Bank study (2005), Hawrylenko (2010), Laura (2010), and Barakat et al. (2013), and Evans and Popova (2015) are just few examples of studies that investigated the important role education can play in post-conflict countries. Almost all such books and studies agreed that "while education systems have the potential to act as a powerful force for peace, reconciliation and conflict prevention by promoting mutual respect, tolerance and critical thinking, they often fuel violence by providing insufficient or unequal access and/or the wrong type of education" (Barakat et al. 2013, p. 125).

Clarke and O'Donoghue (2013) argue that there are important guidelines of the developments of students' achievements that can be attained in schools by enhancing the curriculums, empowering teachers' roles and capacity and enhance school leadership quality. In their book, Kheang et al. (2018) provided for a comprehensive overview of the challenges that face leaders in the education field in developing countries and in post-conflict countries.

It is also important to examine the work of Leithwood and Massey (2010). They believe that "effective leadership" is a major factor for the success of any effort attempting reform. Therefore, there is a great need for promoting assessment skills to improve the academic achievement of students. However, according to Clarke and O'Donoghue (2013), there is very little empirical research that can be relied on in the researcher's attempt to give interpretation to school leadership and assessment management context in post-conflict settings at an international level.

3   Research Methodology and Approach

3.1   Research Methodology

The study will be conducted through both comparative qualitative and comparative quantitative methods. However, the researcher utilized more qualitative tools for the findings and results of the study, basing the study on emerging and current studies and research as well as conducting a case study. The case study was done through several semi-structured in-depth interviews conducted though the telephone with two school principals and two teachers from four Syrian elementary schools, based in areas outside the control of the central government of Syria.

The case study attempted to investigate the awareness and application of PIRLS as a tool of assessment of reading literacy for elementary level students, hoping to answer the major question of this research study: Can PIRLS be tailored and adapted in Syrian elementary schools and utilized as a tool of educational reformation of a broken educational system in the aftermath of conflict or war in Syria? The question of the research was answered based on the statistics and feedback provided by the interviewees. The example reading tests were then provided to the interviewees to submit to their chosen sample of students at 4th grade levels.

3.2   Research Approach

The researcher placed his study within the perspective of both *positivist* and *interpretivist* schools since the study would entail both fixed and flexible underpinnings in reaching the results. First the study was based on the findings and statistics reached to date by prominent authors in the field and internally accredited organizations and establishments, such as IEA and OECD. The researcher based the study on some objectivist findings as well as some subjectivist approach in handling what had already been achieved in the filed so far and what is needed to be done yet.

4 Data Collection and Analysis

Four Syrian elementary schools were chosen for the case study of this paper, located in areas outside the control of the government. Due to the political and social constraints, the interviews were conducted through phone and video calls with two elementary school principals as well as English and Arabic teachers from each school, based in areas that did not possess formal assessment management tools. First, the interviews were conducted to focus on questions related to the existing assessment methods followed by each school and how well they were reflected in the teaching and learning process of fourth grade students and above. Then the questions touched on principals' and teachers' awareness of international assessment measures as well as their application and usage. Finally, the principals and teachers were asked to help in conducting simulated PIRLS assessments to be based on some of the sample reading tests taken from TIMSS and PIRLS resources and some other public resources and websites.

The PIRLS assessment samples that were handed over to fourth grade students in the selected Syrian schools were tailored and adapted to suit the available schooling and educational elements. For the sake of this paper, selected passages from Syrian fourth grade English and Arabic coursebooks were chosen to tailor the assessments. The questions were modeled to suit the selected passages within the Syrian context. All the questions were of the interpretive type, and they were categorized into two sections, each in accordance with the type of selected reading: Reading for Literary Experience and Reading to Acquire and Use and Acquire Information.

4.1 Questions Asked in the Interviews

The questions directed to the interviewee participants are grouped into the following categories:

- *Group 1: About Being a Leader/Teacher in Post-Conflict School Environment*
  Four questions were asked, focusing on the schooling environment of interviewees and how teachers' teaching experience was impacted by the Syrian conflict.
- *Group 2: About Application of Assessment to Assess Students' Achievements*
  Four questions were asked on the kind of assessment measures used by teachers and in which way these measures, if they exist, are controlled by the local authorities.
- *Group 3: About Reading Teaching and Assessment for a 4th Grade Class.*
  Nine questions were asked to the kind of reading materials, textbooks and assessments used for 4th grade students, in both Arabic and English langue contexts.
- *Group 4: About Awareness of PIRLS and How Far it can be Implemented*

Eight questions were asked about interviewees' awareness and knowledge of the PIRLS assessment and if any similar assessments are or have been used. The questions also aimed to obtain information on the kind of support required to enact PIRLS assessment in conflict impacted educational settings.

4.2 Tools for Data Collection and Analysis

Because PIRLS assesses the two major purposes that fourth grade students usually qualify for, Reading for Literary Experience and Reading to Acquire and Use Information, it is important to focus on these two purposes while providing PIRLS exam material as a tool for getting results for this study. The four processes of comprehension to be assessed by PIRLS are widely used by fourth grade student readers, and they "focus on and retrieve explicitly stated information [...] interpret and integrate ideas and information, and evaluate and critique content and textual elements". (PIRLS 2016 Assessment Framework, 2nd Edition). For that purpose, the sample PIRLS passages, questions and scoring guides were an adaptation of the passages posted on TIMSS & PIRLS website: https://timssandpirls.bc.edu/pirls2016/downloads/P16_FW_Appendix_B.pdf.

The reading passages were replaced with reading texts from Syrian fourth grade textbooks, both English and Arabic. The updated and tailored package was sent to the interviewees to carry out the tests on sample students form their respective schools. The scoring guides for the structured assessment questions were sent and explained to the interviewees to make sure they are familiar with the scoring and marking instruments before they submitted the reading passages and questions to their students. After receipt of students' answers and interviewees' scoring cards, the researcher conducted an analysis of the results and examined areas of weaknesses and strengths.

5 Results and Findings and Piloting of the Study

5.1 Results and Findings

The findings show that the best assessment measure that can be adopted in post-conflict and war affected countries and zones is the PIRLS. This mode of assessment is the best choice for education reformation in conflict affected countries. Practices and implementation of PIRLS help build a quality teaching leadership in areas affected by conflict. These indirect effects of the application of enhanced and sustained measures of assessment on high-quality leadership appear to be especially important in schools operating in exceptional and emergency contexts.

However, the researcher found that the reformation of the educational system quality in conflict affected countries such as Syria requires the right programs to be funded and adopted to seek actual education improvements and benefits. It is highly recommended that the PIRLS assessment measure be tailored, first to adapt to the Syrian current emergency and post-conflict factors and second to ensure best quality results for the whole educational process. After the change, an updated and more in-depth version of PIRLS can be adopted and invested on in building the reserve capacity for a smarter and more focused Syrian generation.

An important message from the case study samples was that reform through the application of PIRLS and maybe other international assessment measures in conflict affected areas is feasible. The implementation of tailored PIRLS assessment tools can help create a difference in children's learning and enhances the quality of teaching and learning at large. However, the initiatives that attempt to achieve quality learning in such contexts need to be done on a long-term basis and appropriately investigated. In this sense, a wider scope of academic research needs to be conducted and pursued.

5.2   Discussions of Results and Piloting of the Study

There are at least six implications from both the literature review and the case study findings that can be utilized in the Syrian context. They can be formulated to help support the major questions of a further study at a larger scale within the Syrian and other similar contexts. First, there is an indispensable need for the enhancement of Syrian elementary students' engagement in all kinds of assessment initiatives, most importantly international assessment measures. Second, it has been found that there is no effective and capable substitute for the formal and official educational system in the country. Third, through the application of PIRLS, there is a lot to be gained from the innovations of advanced countries and some developed countries. Fourth, there is no magical solution when it comes to the adoption and implementation of a comprehensive assessment approach in post-conflict countries. Fifth, any attempt at educational reform to improve student learning is of marginal use unless there are systems in place to hold the people who are performing the reform accountable. Finally, it has been found that the implementation of solid assessment and learning tools in post-conflict countries requires investing on capabilities and technology skills and advanced educational tools. As a result, a further study will focus more on the qualifications and skills that are required for the improvement of learning quality and empowering teachers as leaders in communities affected by war and conflict.

References

Barakat, S., Connolly, D., Hardman, F., & Sundaram, V. (2013). The role of basic education in post-conflict recovery. *Comp. Educ., 49*(2), 124–142.

Bernbaum, B. & Moore, A. S. (2016). *Examining the Role of International Achievement Tests in Education Policy Reform: National Education Reform and Student Learning in Five Countries*, A Policy Paper published by USAID, EQUIP2.

Clarke, S., & O'Donoghue, T. (2013). The case for studying educational leadership at the individual school level in post-conflict societies. In Clarke, S. & O'Donoghue, T. (Eds.),*School level leadership in post-conflict societies: The importance of context* (pp. 1–8). London/New York: Routledge.

Davies, L. (2004a). Building a civic culture post-conflict. *Lond. Rev. Educ., 2*(3), 229–244.

Davies, L. (2004b).*Education and conflict: complexity and chaos.* London: Routledge

Evans, D. & Popova, A. (2015). What really works to improve learning in developing countries? An analysis of divergent findings in systematic reviews. *Policy Research Working Papers.* WPS7203, Washington DC: World Bank.

Gallagher, T. (2004). *Education in divided societies.* New York: Palgrave Macmillan

Harlen, W. (2004). *A systematic review of the evidence of impact on students, teachers and the curriculum of the process of using assessment by teachers for summative purposes.* London Institute of Education: EPPI-Centre. http://eppi.ioe.ac.uk/EPPIWeb/home.aspx?page=/reel/review_groups/assessment/review_four.htm

Hawrylenko, J. (2010). *Education in post-conflict societies.* Alberta: Athabasca University

James, M. (2006). Assessment, teaching and theories of learning.*Chapter 3 in J. Gardner (Ed) (2006) Assessment and Learning,* (pp 47–60. First Ed.). London: Sage.

Jones, B., Wittes, T. C. & Yerkes, S. (2017).*Stabilization planning in Syria.* Washington: Brookings.

Kheang, T., O'Donoghue, T., & Clarke, S. (2018). Educational leadership in developing countries and in post-new war countries. In *Primary School Leadership in Cambodia: Context-Bound Teaching and Leading.* pp. 43–84.

Leithwood, K., & Massey, L. (2010). *Developing Leadership to improve Student Outcomes.*,. https://doi.org/10.1007/978-90-481-9106-2_6.

Shi, W., He, X., Wang, Y., Fan, Z., & Guo, L. (2016). PISA and TIMSS science score, which clock is more accurate to indicate national science and technology competitiveness? *Eurasia J. Math. Sci. Technol. Educ., 12*(4), 965–974.

Sommers, M. (2002). Children, Education and War: Reaching Education For All (EFA) Objectives in Countries Affected by Conflict. *Conflict Prevention and Reconstruction Unit Working Papers.* Paper No. 1, pp. 1–40.

The Assessment Department of the UAE Ministry of Education. (2012). *Building Knowledge in the UAE Report, Findings from TIMSS 2011 & PIRLS 2011.* UAE Ministry of Education Website.

The International Rescue Committee (IRC). (2017). *Impact of war on Syrian children's learning* Report.

Using Nearpod to Promote Engagement in Online ESL Classes: A Mixed-Methods Study in the Context of Higher Education

Azza Alawadhi and Rawy A. Thabet

Abstract Student Response Systems such as *Kahoot!*, *Socrative* and *Nearpod* have become one of the latest trends in teaching and learning across higher education. However, despite the popularity of these platforms, the integration of SRS in teaching is still an evolving field of study. This mixed-methods study draws on undergraduate students' perceptions of using Nearpod to facilitate teaching and learning in an online English course at a federal higher education institution in the UAE during pandemic teaching. A combination of self-report surveys ($N = 90$) and in-depth interviews ($N = 5$) were used to collect data for this study. Findings suggest that students perceived Nearpod to promote fun and enjoyment, enhance knowledge and understanding, and improve classroom dynamics. Results indicate a generally positive response, with 93.3% of students reporting that the instant feedback afforded by Nearpod improved their understanding, while 83.4% reported an increase in interactivity. This study confirms previous findings, suggesting that SRS such as Nearpod could foster effective student engagement, increase participation, and enhance students' online learning experience. The study also found that there were no significant gender differences in students' perceptions of Nearpod. Pedagogical implications are further discussed, and future research suggestions are provided.

Keywords Student response system · Nearpod · Engagement · Active learning · Mixed-methods

1 Introduction

The last few years have seen an increased interest in the use of *Student Response Systems* (SRS) in higher education (Hassan et al., 2021). However, much of the published studies focused on K-12 education (Wang & Tahir, 2020), particularly on Kahoot! (Wang, 2015; Plump & LaRosa, 2017), and Socrative (Coca & Slisko, 2013; Awedh et al., 2014), mainly using experiments. In addition, evidence from

A. Alawadhi (✉) · R. A. Thabet
The British University in Dubai, Dubai, UAE
e-mail: aalawadhi@hct.ac.ae

© The Author(s) 2023
K. Al Marri et al. (eds.), *BUiD Doctoral Research Conference 2022*,
Lecture Notes in Civil Engineering 320,
https://doi.org/10.1007/978-3-031-27462-6_11

117

previous studies was based on data generated in face-to-face settings, while little research addressed undergraduate students' perceptions of SRS during online classes (Holbrey, 2020). Although Nearpod has been a popular tool at universities and colleges in the UAE for a while, the report on these practices remains largely unexplored. Therefore, this paper explores how Nearpod, a *Student Response System* (SRS) has been used in online English classes to improve engagement and facilitate teaching during the pandemic. In the literature, these interactive platforms have been identified under a variety of names. For example, Caldwell (2007) uses the term clickers, while Kay and LeSage (2009) refer to these platforms as Audience-response Systems. Likewise, Wang (2015) uses the term Student Response System. In this paper, we refer to these interactive platforms as SRS.

Nearpod is a free web-based platform that allows instructors to create engaging lessons and interactive presentations embedded with multimedia, videos, virtual reality tours, and self-paced quizzes (Fig. 1). Nearpod provides real-time formative assessment supported with instant feedback, which enables students to reflect on their learning immediately. Nearpod also helps instructors track students' progress, identifying areas where students need more practice and highlighting them instantly (Fig. 2). As compared to other SRS, Nearpod offers additional opportunities for real-time interactions beyond the traditional MCQ's including live polls, voice-recording feature, drawing function, gamification and competition through the use of 'Time to Climb' game. Furthermore, instructors can identify students who remain active and monitor their progress and attendance. Despite the many uses of this platform, this study focuses on using Nearpod to facilitate teaching and learning English in online lessons.

Fig. 1 Question types in Nearpod

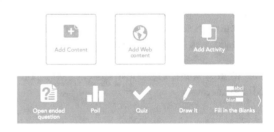

Fig. 2 Students' progress report

1.1 Research Aim and Questions

Considering the aforementioned background, this mixed-methods study aims to investigate undergraduate students' perceptions of Nearpod, an SRS that is used to facilitate active learning in an English language online course at a federal higher education institution in the UAE during pandemic teaching. The following research questions guided the study:

> **RQ1:** How do undergraduate students perceive the use of Nearpod for teaching and learning during online ESL classes?
> **RQ2:** Are there gender differences in undergraduate students' perceptions of Nearpod?

2 Theoretical Perspectives

Active learning is generally defined as a student-centered pedagogical approach that helps to promote interaction, collaboration, discussion, and self-reflection. Prior research (Armbruster et al., 2009; Shroff et al., 2019) suggests that effective integration of active learning strategies may improve students' attitudes toward learning, maximize cognitive engagement and assist faculty in achieving successful learning outcomes. Moreover, studies have demonstrated that active learning strategies such as timely feedback may enhance students' self-efficacy and improve retention of knowledge. Student engagement is a defining construct of active learning. Engagement is often cited as students' interest, active interaction, involvement, and participation in purposeful learning activities which enhance learning performance to achieve learning outcomes (Fredricks et al., 2004). Research shows (Blasco-Arcas et al., 2013) when students are engaged in their learning, they demonstrate increased attention, motivation, participation, and ultimately show high levels of satisfaction, and enjoyment, which are directly linked to student success.

3 Literature Review

Considerable scholarly attention has been given to the benefits of using SRS in the classroom (Wang & Tahir, 2020; Kocak, 2022). Most of these studies have helped identify the key advantages of using SRS to enhance teaching and learning in traditional face-to-face instruction (Wang, 2015; Göksün & Gürsoy's, 2019). For instance, the seminal work of Caldwell (2007) shows that these interactive technologies positively impacted classroom dynamics, stimulated class discussion, increased learning effectiveness, and improved attendance. More recently, Holbrey (2020) studied Kahoot!, a similar SRS tool, and found that it was a successful platform in enabling active participation. Among the other benefits, SRS has been found to provide meaningful and timely feedback. Other studies (Hung, 2017; Hassan et al., 2021) have shown that instant feedback in the form of scores, badges, rankings, and rewards has led to higher student engagement. Research also shows that SRS reduces classroom anxiety through anonymous participation (Wang, 2015). In a similar study, Plump and LaRosa's (2017) survey results reported that students perceived SRS, namely Kahoot!, to improve their comprehension, facilitate learning new concepts, and make learning more enjoyable. Consistent with Plump and LaRosa's (2017) findings, Shehata et al. (2020) verified that SRS may improve students' learning outcomes. In a similar vein, Dizon (2016) found evidence that SRS (Quizlet) was effective in EFL students' vocabulary acquisition. However, not all studies share these positive views of SRS. Research by Kaya and Balta (2016) demonstrated that although SRS (Socrative) stimulated more interactive discussions, no difference was noted in students' academic outcomes.

In reviewing the literature, studies revealed that students perceive SRS to be beneficial, although evidence of improved learning has been less clear and inconsistent (Coca & Slisko, 2013; Kaya & Balta, 2016). Apart from Hakami (2020) study, there is a general lack of research from the gulf context documenting the use of Nearpod in online teaching during the pandemic. To the authors' knowledge, no study has formally investigated Emirati students' perceptions of Nearpod during online teaching across tertiary education in the UAE. In an attempt to address the gap, this mixed-methods study aims to investigate undergraduate students' perceptions of using Nearpod to facilitate teaching and learning in an ESL program at a federal higher education institution in the UAE.

4 Research Methods

4.1 Research Design

This study adopted a mixed-methods (QUAN + QUAL) concurrent triangulation design (Creswell, 2007), where both quantitative and qualitative data were collected in parallel in a single study (Fig. 3). A survey ($N = 90$) was administered to find out undergraduate students perceptions of using Nearpod in online lectures, alongside in-depth interviews ($N = 5$). According to Creswell (2007), triangulation adds rigor, depth, and trustworthiness to the study.

4.2 Sample and Setting

This small-scale study was conducted in the Spring semester of 2021 in an undergraduate ESL program at a medium-sized federal higher education institution in the UAE. A convenience sample of ninety ($N = 90$) participants completed the survey.

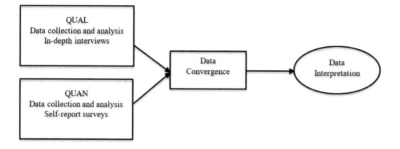

Fig. 3 Concurrent triangulation design adapted from Creswell (2007)

Table 1 Participant's demographic information

Demographic	Frequency	Percentage
Age	(n)	(%)
17–18	54	60
19–20	32	35.6
21–22	1	1.1
Above 22	4	4.4
Gender		
Male	28	31.1
Female	62	68.9
(N = 90)		

Five female ($N = 5$) students volunteered to participate in the in-depth interviews. Table 1 provides the general profile of the participants.

4.3 Data Collection Instruments and Analysis

A survey was adapted from Shehata et al. (2020) to capture students' perceptions of using Nearpod in online ESL classes and ensure the content validity of the survey. The survey was administrated using Google Forms. The first part of the survey elicited students' demographic information. The second part of the survey included 10-items which were rated on a 5-point Likert scale ranging from 5 (*Strongly Agree*) to 1 (*Strongly Disagree*). Cronbach's alpha was ($\alpha = .873$), indicating a good internal consistency.

In-depth online interviews were conducted via *Blackboard Collaborate Ultra* video conferencing after obtaining informed consent. The interviews ($N = 5$) consisted of seven open-ended questions that focused on understanding students' experiences with Nearpod; and how it affected their virtual learning experience. The interviews lasted approximately 20–30 min. The interviews were recorded and transcribed for analysis.

Table 2 Students' perceptions of using Nearpod in the online classroom

Items	Strongly agree (5)	Agree (4)	Neutral (3)	Disagree (2)	Strongly disagree (1)	SD	Mean
Item (1) I enjoy using Nearpod in my English lessons	50%	25.6%	15.6%	7.8%	1.1%	1.027	4.16
Item (2) Nearpod helps me understand the English lesson better	62.2%	16.7%	16.7%	3.3%	1.1%	.952	4.36
Item (3) I think that Nearpod helps shy students interact more	61.1%	16.7%	20%	0%	2.2%	.950	4.34
Item (4) Nearpod creates an interactive learning environment	55.6%	27.8%	13.3%	3.3%	0%	.839	4.36
Item (5) The level of interactivity in the class increased with the use of Nearpod	61.1%	21.1%	14.4%	3.3%	0%	.859	4.40
Item (6) Using the Nearpod's interactive elements increased my understanding	51.1%	26.7%	15.6%	5.6%	1.1%	.977	4.21
Item (7) Doing the multiple choice questions helped me to understand things better	55.6%	26.7%	13.3%	4.4%	0%	.874	4.33
Item (8) There is no difference in teaching with or without using Nearpod	26.7%	22.2%	23.3%	17.8%	10%	1.320	3.38

(continued)

Table 2 (continued)

Items	Strongly agree (5)	Agree (4)	Neutral (3)	Disagree (2)	Strongly disagree (1)	SD	Mean
Item (9) I enjoy learning when using the Nearpod integrated with the video	56.7%	22.2%	17.8%	2.2%	1.1%	.920	4.31
Item (10) I like to receive instant feedback using Nearpod	78.9%	14.4%	3.3%	2.2%	1.1%	.747	4.68

Total items 10
(N) 90

Descriptive statistics were calculated using SPSS version 23 to provide the mean and standard deviation for the survey items. *T-test* was used to determine if there are any gender differences in undergraduate students' perceptions of Nearpod. Moreover, the data from the Likert-scale items were analyzed using Google Forms which provided figures and percentages. The qualitative data obtained from the in-depth interviews were analyzed based on the theoretical positions of Braun and Clarke's (2006) six-phase thematic analysis. Trustworthiness was achieved from repeated checking of the data and verifying internal data consistency by multiple researchers.

5 Findings and Discussion

This section is structured based on the research design. First, quantitative data are presented; then, the qualitative findings are analyzed and discussed.

5.1 Quantitative Results

RQ1: How do undergraduate students perceive the use of Nearpod for teaching and learning during online ESL classes?

Table 2 shows the percentages and descriptive statistics from the survey. Interestingly, most of the items were rated positively. Almost 76% of those surveyed enjoyed using Nearpod in English lessons. It is notable that over 78% of the respondents agreed that Nearpod helped shy students interact more. This finding indicates that students feel more comfortable participating in online platforms, which reaffirms the research findings of Caldwell (2007) and Wang (2015), who acknowledge that students appreciate the anonymous participation afforded by SRS. They argue that when students believe their responses are anonymous, their active participation increases. In addition, about 83% believed that Nearpod creates an interactive learning environment. These results correspond with Blasco-Arcas et al. (2013) who concurred that promoting engagement and interactivity is critical as it leads to better and more effective learning. Of the 90 students who used Nearpod, approximately 82.2% perceived Nearpod to increase the level of interactivity. The data also demonstrates that about 78% believed that Nearpod's interactive elements increased their understanding of the lesson. In addition, more than 82.3% of those surveyed reported that doing the multiple-choice questions helped them understand things better. These findings related to improved learning and understanding align with the research conducted by Plump and LaRosa (2017) and Göksün and Gürsoy's (2019), who found that SRS improved students' knowledge retention and learning outcomes. As shown in Fig. 4, nearly 79% of the respondents felt that they enjoyed learning when using Nearpod integrated with the video. Interestingly, as shown in (Fig. 5), the statement 'I like to receive instant feedback using Nearpod' scored positively at 93.3%. It appears that students perceive that the instant feedback afforded by Nearpod promotes deep learning, which supports the previous work of Shroff et al. (2019) on the value of instant feedback afforded by SRS.

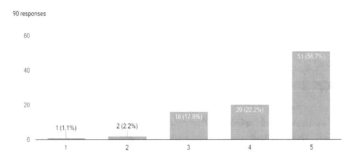

Fig. 4 I enjoy learning when using Nearpod integrated with the video

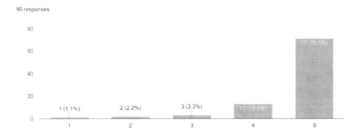

Fig. 5 I like to receive instant feedback using Nearpod

Table 3 Difference between male and female students

	Gender	N	Mean	Std. deviation	Std. error mean
Perception	Male	28	4.3179	.67772	.12808
	Female	62	4.2226	.64641	.08209

RQ2: Are there gender differences in undergraduate students' perceptions towards Nearpod?

As shown in Table 3, there was a slight difference between the means of male and female students in terms of the level of agreement towards Nearpod. Table 4 shows no statistically significant difference between male and female students ($t = .638$, $df = 88$, $p = .525$, two-tailed).

Table 4 shows no statistically significant difference between male and female students ($t = .638$, $df = 88$, $p = .525$, two-tailed).

5.2 Qualitative Results

Theme One: Fun and Enjoyment

All of the participants ($N = 5$) seem to comment favorably on the engagement afforded by Nearpod. Notably, undergraduate students described their experience with Nearpod as 'motivating,' 'interesting,' and 'engaging'. In addition, 'fun,' 'enjoying,' and 'exciting' were some of the most repeated phrases during the interviews. Students also perceived Nearpod to make class time 'more productive' and felt 'less bored' with the use of Nearpod. These findings from the interviews are in line with previous studies (Hung, 2017; Hassan et al., 2021), which confirmed that SRS has proven to affect students' acceptance of the lesson, promote entertainment, and foster a positive atmosphere.

Theme Two: Knowledge and Understanding

Some participants noted that Nearpod provided them with learning guidance, especially in online lectures. Many students also mentioned that Nearpod allowed them to be 'active' in class, 'collaborate more', and 'increased their understanding'. These findings are similar to past studies on SRS which have given special consideration to its impact on enhancing understanding and knowledge retention (Coca & Slisko 2013; Kaya & Balta's, 2016; Plump & LaRosa, 2017). Some participants mentioned that using Nearpod for teaching new vocabulary and grammar was 'useful' and gave them a chance to 'discuss the correct answer'. Comparable results were also obtained by Dizon (2016), who report that SRS improves students' retention of vocabulary in EFL classrooms.

> "Nearpod was useful to revise vocabulary and grammar" (Participant A).

> "To receive the instant feedback was very good to know my mistakes immediately" (Participant E).

Table 4 Independent sample T-test for gender differences

		Levene's Test for Equality of Variances		t-test for Equality of Means						95% confidence interval of the difference	
		F	Sig.	t	df	Sig. (2-tailed)	Mean difference	Std. error difference		Lower	Upper
Perception	Equal variances assumed	.144	.705	.638	88	.525	.09528	.14940		−.20163	.39219
	Equal variances not assumed			.626	50.007	.534	.09528	.15213		−.21028	.40083

Students also appreciated the interactive functions afforded by Nearpod, such as live polls, voice-recorder, and drawing function, which made learning more pleasurable. These findings align with previous studies conducted by Wang and Tahir (2020), who confirmed that instant feedback, scores, and leader boards are some of the most useful features provided by SRS appreciated by students. Furthermore, all participants referred to Nearpod as an effective tool to 'provide immediate feedback', which helped them learn 'quicker'. Some participants also reported that tracking their progress was more efficient online, as they could quickly identify their mistakes in real-time. These findings follow Awedh et al., (2014) research findings, who stated that SRS is a good review tool to keep students focused and engage them in active learning. Some examples of participants' comments follow:

"Nearpod helps me learn my mistakes quickly" (Participant A).

"It was interesting to see my results and know if I understood the content" (Participant E).

"I interact more with the teacher and my classmates when I play Nearpod" (Participant B).

Theme Three: Dynamics of the Lesson

Qualitative data analysis identified issues related to the dynamics of online lessons. Almost all of the participants ($N = 5$) were in favor of SRS, indicating that Nearpod made their learning more 'personalized', 'active' and gave them a 'break' during the long hours spent on the screen. Students were also excited to try new features of Nearpod, such as virtual tours and live polls that seem to increase their satisfaction of the lesson. These findings match those observed in earlier studies of Wang (2015) who sought to demonstrate the benefits of SRS in differentiating instruction, increasing participation, and improving the classroom dynamics. Many participants recognized that they needed more engaging platforms such as Nearpod and Kahoot! to stimulate their interest in the lesson, keep them 'awake' and 'focused.'

The current study demonstrated that both the quantitative and qualitative data indicated that students see SRS as a promising tool to foster engagement and enhance interactivity in online classes. Overall, the quantitative and qualitative data suggest that students generally perceive Nearpod as useful in online ESL lectures. In addition, it appears that students are likely to spend more time in online classes if it is enjoyable, engaging, and involve gamification. These findings reinforces the pedagogical characteristics of SRS as cited in existing literature, including the work of Alawadhi and Abu-Ayyash (2021); Holbrey (2020), and Göksün and Gürsoy (2019).

6   Limitations and Further Research

With reference to limitations, the following should be considered. The sample population has been drawn from a single college, therefore, the results cannot be generalized. In addition, the results are based on students' self-report data, which is subject to social desirability bias (Johnson & Christensen, 2014). Nonetheless, this study has significant implications for theory and practice in ESL tertiary education. It is hoped that this study will provide college instructors with valuable insights for addressing the problem of low engagement in online classes. Instructors could benefit from the availability of SRS to improve teaching practices in online classes to support more active participation and better engagement.

7 Conclusion

This study investigated undergraduate students' perceptions of using Nearpod, an SRS to facilitate teaching and learning in an English language course at a federal higher education in the UAE. An online survey ($N = 90$) and in-depth interviews ($N = 5$) were conducted to collect data for this study. The findings suggest that ESL students perceive the use of Nearpod as an effective tool to enhance knowledge and understanding and improve classroom dynamics. This study also confirms that SRS is emerging as a promising tool to foster engagement in online learning. Though COVID-19 impacted conventional teaching and learning, universities and colleges may take this as an opportunity to reform their teaching practices through emerging new technologies. As in the UAE tertiary institution landscape, more scholarly work is needed in this field. Therefore, the findings of this study may contribute to our understanding of the role of SRS in online learning during pandemic teaching.

References

Alawadhi, A., & Abu-Ayyash, E. A. (2021). Students' perceptions of Kahoot!: An exploratory mixed-method study in EFL undergraduate classrooms in the UAE. *Education and Information Technologies, 26*(4), 3629–3658.

Armbruster, P., Patel, M., Johnson, E., & Weiss, M. (2009). Active learning and student-centered pedagogy improve student attitudes and performance in introductory biology. *CBE—Life Sciences Education, 8*(3), 203–213.

Awedh, M., Mueen, A., Zafar, B., & Manzoor, U. (2014). Using socrative and smartphones for the support of collaborative learning. *International Journal on Integrating Technology in Education, 3*(4), 17–24.

Blasco-Arcas, L., Buil, I., Hernández-Ortega, B., & Sese, F. J. (2013). Using clickers in class. The role of interactivity, active collaborative learning and engagement in learning performance. *Computers & Education, 62*, 102–110.

Braun, V., & Clarke, V. (2006). Using thematic analysis in psychology. *Qualitative Research in Psychology, 3*(2), 77–101.

Caldwell, J. E. (2007). Clickers in the large classroom: current research and best-practice tips. *CBE—Life Sciences Education, 6*(1), 9–20.

Coca, D. M., & Slíško, J. (2013). Software Socrative and smartphones as tools for implementation of basic processes of active physics learning in classroom: an initial feasibility study with prospective teachers. *European Journal of Physics Education, 4*(2), 17–24.

Creswell, J. W. (2007). *Research design: qualitative, quantitative, and mixed methods approaches* (3rd ed.). Sage Publications, Inc.

Dizon, G. (2016). Quizlet in the EFL classroom: enhancing academic vocabulary acquisition of Japanese university students. *Teaching English with Technology, 16*(2), 40–56.

Fredricks, J. A., Blumenfeld, P. C., & Paris, A. H. (2004). School engagement: potential of the concept, state of the evidence. *Review of Educational Research, 74*(1), 59–109.

Göksün, D. O., & Gürsoy, G. (2019). Comparing success and engagement in gamified learning experiences via Kahoot and Quizizz. *Computers & Education, 135*, 15–29.

Hakami, M. (2020). Using nearpod as a tool to promote active learning in higher education in a BYOD learning environment. *Journal of Education and Learning, 9*(1), 119–126.

Hassan, M. A., Habiba, U., Majeed, F., & Shoaib, M. (2021). Adaptive gamification in e-learning based on students' learning styles. *Interactive Learning Environments, 29*(4), 545–565.

Holbrey, C. E. (2020). Kahoot! Using a game-based approach to blended learning to support effective learning environments and student engagement in traditional lecture theatres. *Technology, Pedagogy and Education, 29*(2), 191–202.

Hung, H. T. (2017). Clickers in the flipped classroom: bring your own device (BYOD) to promote student learning. *Interactive Learning Environments, 25*(8), 983–995.

Johnson, B., & Christensen, L. (2014). *Educational research: quantitative, qualitative, and mixed approaches* (5th ed.). SAGE publications.

Kay, R. H., & LeSage, A. (2009). Examining the benefits and challenges of using audience response systems: a review of the literature. *Computers & Education, 53*(3), 819–827.

Kaya, A., & Balta, N. (2016). Taking advantages of technologies: using the Socrative in English language teaching classes. *International Journal of Social Sciences & Educational Studies, 2*(3), 4–12.

Kocak, O. (2022). A systematic literature review of web-based student response systems: advantages and challenges. *Education and Information Technologies, 27*(2), 2771–2805.

Plump, C. M., & LaRosa, J. (2017). Using Kahoot! in the classroom to create engagement and active learning: a game-based technology solution for eLearning novices. *Management Teaching Review, 2*(2), 151–158.

Shehata, N., Mitry, C., Shawki, M., & El-Helaly, M. (2020). Incorporating Nearpod in undergraduate financial accounting classes in Egypt. *Accounting Education, 29*(2), 137–152.

Shroff, R. H., Ting, F. S., & Lam, W. H. (2019). Development and validation of an instrument to measure students' perceptions of technology-enabled active learning. *Australasian Journal of Educational Technology, 35*(4).

Wang, A. I. (2015). The wear out effect of a game-based student response system. *Computers & Education, 82*, 217–227.

Wang, A. I., & Tahir, R. (2020). The effect of using Kahoot! for learning – a literature review. *Computers & Education, 149*, 103818.

Investigating Emirati Students' Practices of Learning Block-Based Programming in an Online Learning Context: Evidence from a UAE National Program

Wafaa Elsawah and Tendai Charles

Abstract Purpose The primary purpose of this study is to investigate Emirati students' perceptions about a two-week online programming course.

Methodology A mixed-method approach was used to report a two-week program experience in which 913 Emirati students engaged in a programming course. During the program, students' progress and achievements were observed. At the end of the program, a survey was distributed amongst students for them to report on their perceptions about the experience.

Findings - Results revealed that most of the students enjoyed learning programming in an online environment and they fully comprehended the newly presented concepts; however, a few of them experienced various forms of technical difficulties.

Implications - This study contributes to the growing body of literature on the value of programming skills and the role of online learning environments in developing these skills for school students, especially for students in the United Arab Emirates (UAE).

Originality - The outcomes of this study may assist educational policymakers in the UAE to enhance their implementation of online learning, particularly in programming. Moreover, it may help educators better prepare for the problems that students face with online programming.

Keywords Programming · Coding · Computational thinking · Tynker · Online learning

1 Introduction

Throughout the past decade, programming has been identified as a valuable skill that young students should attempt to acquire in order to meet the demands of a 21st-century workforce (Abiodun & Lekan, 2020; Vico et al., 2019). In the same

W. Elsawah (✉) · T. Charles
The British University in Dubai, Dubai, UAE
e-mail: 20000587@student.buid.ac.ae

© The Author(s) 2023
K. Al Marri et al. (eds.), *BUiD Doctoral Research Conference 2022*,
Lecture Notes in Civil Engineering 320,
https://doi.org/10.1007/978-3-031-27462-6_12

vein, the paradigm shift that the educational sector experienced during the COVID-19 pandemic led many schools to reorient their pedagogical strategies by adopting 'online teaching and learning'. Under these rapid changes, new opportunities to learn programming were presented, which emphasised the design and delivery of compelling learning experiences. This study focuses on a United Arab Emirates (UAE) national program to investigate the students' practises of learning block-based programming, which is based on visual/graphical ways that matches the students' age. The program was launched by a governmental organisation to teach Emirati children aged 7–14 the basics of programming, with the intentions of creating a new generation that is skilled with contemporary programming language skills.

With the wide spread of technology, many children are tech-savvy from a young age; however, they do not understand the true technical workings of computer technology (Kaplancali & Demirkol, 2017). Additionally, one could argue that most UAE schools do not allocate adequate material and time to teaching computing concepts and skills. This may be due to the priority given to other subjects, like math and science. Therefore, teachers could take advantage of the spread of online learning and direct it in a way to build students' knowledge in computing. In essence, this study attempts to highlight the critical need for programming skills for K-8 students and then investigate the effectiveness of using online learning environments to meet this need.

Investigating students' practices in online learning environments is one of the current trends that many studies have discussed. However, inadequate attention has been paid to teaching programming to K-8 children (Lewis, 2020; Vico et al., 2019), especially in the online learning context in the UAE. It seems school leaders in the region are unaware of the fact that programming skills are crucial for all students and can be taught using different pedagogical approaches. Therefore, the current study attempts to address this gap in the literature. This paper highlights the importance of programming skills for K-8 students essential to cope with 21st requirements and investigates the students' practice of learning programming in an online learning classroom, taking into account the students' levels of understanding. The purpose of this paper, therefore, is driven by attempts to answer the following questions:

- How effective are online learning environments for acquiring programming skills?
- What are the students' reflective perceptions of the online programming course?

2   Literature Review

The Technology Acceptance Model (TAM) proposed by Davis (1989) was adopted as a theoretical framework for this study. TAM focuses on two major elements that affect an individual's willingness to adopt new technology: perceivable ease of usage and perceivable usefulness (Boot & Charness, 2016). Students who perceive coding as too complex to learn, especially in such online learning environments, will be unlikely to learn it. Chao (2016) and Kong et al. (2018) argue that students may lack enthusiasm and confidence in programming learning because of the belief that

Fig. 1 Technology acceptance model (Davis, 1989)

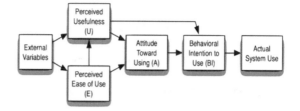

programming requires sophisticated concepts. In contrast, students who perceive coding as providing needed future skills and as easy to understand will be more likely to learn (Fig. 1).

Papert (1972, 1980) also holds the view that children are programming the computer rather than computers programming the children in his constructionism theory that has been utilized and reported extensively in computing and technology literature. He believed that school children possess the capacity to learn and develop a deep understanding of programming, if they start learning from a young age. Recently, researchers have explored the adoption of online coding platforms, which may be especially well-suited for school students (Manita et al. 2021; Stephany et al. 2021). These platforms make coding simple for children to grasp; they rely on visual coding, which gamifies activities, uses goals, stories, and discoveries, and provides a more graphically appealing environment. Gray and Thomsen (2021) reveal in a recent study that the students who learn coding through digital play and playful approaches rapidly engage in the problem-solving process and make significant discoveries.

Programming is also a challenging subject for teachers to teach, and little attention has been paid to the teaching of programming in an online learning context (Nariman, 2021; Watson & Li, 2014). A recent study by Galdo et al. (2022) highlighted the fact that online education provides significant difficulties for students when it comes to programming. However, other researchers draw attention to the distinctive benefits of learning programming in an online context. These benefits include evaluating the students' solutions in real-time and providing instant feedback (Daradoumis et al., 2019). Some authors (e.g., Abiodun & Leka, 2020) have attempted to compare the level of knowledge obtained during an online coding program versus that acquired via traditional methods coding. Despite the findings indicating that both interventions enhanced students' computational thinking skills and competency, the class activities were limited in the online learning context by difficulty in grouping participants and technical problems caused by the Internet and power outages.

A growing body of literature has investigated the effectiveness of teaching block-based programming, especially for younger students. Tsai (2019) explored students' perceptions after learning block-based programming and noted that the students found visual programming to be a helpful tool since they could look up relevant material on the Internet and learn from their mistakes. Furthermore, teaching block-based programming may considerably reduce student dropout rates (Benotti et al., 2018) and protect young students from making programming syntax errors (Ouahbi et al., 2015). However, some authors questioned its usefulness, claiming that when students transition to text-based coding, they feel overwhelmed by the language's structure (De & Do, 2021; Lewis, 2020). Yet, Stead and Blackwell (2014) reported that block syntax could be viewed concurrently with text syntax in most online coding environments, and proven testing with students aged 11–12 have demonstrated that beginning with blocks rather than text improves children's comprehension of text syntax. Together, these studies demonstrate that programming is crucial for all students; nevertheless, some conflicting findings imply that additional research is needed to determine the effectiveness of programming, particularly block-based coding, in an online learning context.

3 Methodology

This paper utilises a mixed-method approach, which includes collecting qualitative data from classroom observation and quantitative data collected from surveys. The mixed-method approach is ideal since it combines the advantages of both (Creswell, 2018). Observation is an appropriate method for this study purpose because it allows the researcher to explore the detailed learning experience of the students in programming through an online environment. The survey aims to validate the qualitative results and avoid the bias that may occur from the researcher as she observes and represents her own work. The sample consists of 913 Emirati students from different schools and Emirates aged 7 to 14. The students were placed into two categories according to their age, the first category had younger students, and the second one had older students. All students had no prior experience with programming and were not familiar with the platform (Table 1).

The program included curriculums from Tynker coding platform. Programming 201 was selected for the younger category and programming 301 for the older one. The curriculum is comprised of nine lessons that include unplugged exercises that teach the students how to work on computational thinking problems and Python programming challenges by creating animated stories, and games (Table 2).

Students' performance was primarily observed in the first phase of data collection, and students' lesson progress was considered and added to the observation results. In the second phase, an anonymous survey with close-ended questions was created using SurveyMonkey. The answers measure the participants' perceptions about learning programming in an online learning environment. To ensure the study's

Table 1 Demographic data of participants		Younger students	Older students	Trainers
	Number of participants	584	329	21
	Age	7–10 years old	11–14 years old	24–40 years old

Table 2 Practical lessons involved in the program for younger and older students

Practical lessons	Programming 201	Programming 301
P1	Introduction to programming 201	Introduction to programming 301
P2	Loops and animation	Motion and tracking
P3	Creating a scene	Conditional loops
P4	Jumping over obstacles	Nested loops
P5	Storytelling	Messaging
P6	Rotation	Shoot projectiles
P7	Broadcasting messages	Cloning
P8	Time limits	Variables
P9	Pop the balloon	Boss Battle

validity, students who respond with a strong agreement or agreement were counted, additionally, the survey has both positive and negative items to reduce the effects of bias.

This program switched to online learning in response to the COVID-19 pandemic in March 2020. As a result, changes have been made to provide the program in online settings. The goal was to prepare students to do the programming tasks remotely, familiarize them with the coding platform (Tynker), and introduce them to basic programming concepts. To do so, we sent them an introductory video that showed them the importance of the program and the basic functionalities that they would need to use during the program two days before the program. A pilot lesson was sent to them one day before the program to ensure that the students were ready for the first day and not have any problem logging into the platform. The program's sessions were organized using Zoom and WhatsApp groups were used as a way of communication between students and trainers. During the program, the students had to attend one session daily in which the trainers explain all the new concepts of the lesson; after that, the students were heading to Tynker to start working on their lesson based on what they understood from the session. The trainers were following up with all students to make sure that they are working properly and help them overtake any technical difficulty. The program has been developed in accordance with the constructionist theory and its basic principle.

Table 3 Percentage of students' attendance, progress, and withdrawal

	7–10 Age Category	11–14 Age category
Withdrawal %	7%	11%
Attendance %	95%	86%
Progress %	97%	89%

4 Data Analysis and Results

Students faced some technical difficulties that impaired their ability to work on the first day. Some of them were confused about the process of activating their Tynker accounts and accessing their class to begin working. However, with the trainers' support and following the videos' instructions, they had begun immersing and working by the end of the day. Regarding the Tynker platform, it allowed students to engage with the tasks in a setting comparable to a real-world environment. Furthermore, Tynker helped teachers collect data on students' participation and development and track nearly every student's action. Students' quiz results indicate that most students completed assigned lessons and made significant progress toward learning to code; on the other hand, a few students fell behind because of the lack of passion in the online learning environment. As a result, these students withdrew due to their inability to navigate the platform and comprehend the coding fundamentals. Surprisingly, the results showed that the withdrawn students from the older category are more than the withdrawn younger ones; moreover, the younger students attended and achieved more progress during the program than the older ones (Table 3).

Five hundred eleven (511) responses were collected from the students, and frequencies and percentages for each question were analysed. The results showed that most students were satisfied with the programming learning experience and eager to join more online programs in the future. The data collected was extracted in PIVOT tables which provided visual charts and analysed accordingly. The first question of the survey measured the students' overall evaluation on learning to program in the online classroom. The results showed that most of the students are very satisfied with their learning experience (Fig. 2).

The next two questions were about the Tynker platform, how the students found it, and the frequency of using it (Figs. 3 and 4).

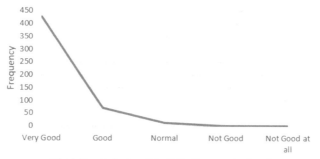

Fig. 2 Evaluation of the online programming learning experience

Fig. 3 Frequency of students entering the platform

'How many times have you entered Tynker platform and completed the lessons through it?': Daily accounts for the majority of 'Frequency'.

Fig. 4 The ease of learning in the online Tynker

'How did you find online learning through Tynker platform?': Easy accounts for the majority of 'Frequency'.

The following part was about the trainers' support. Most of students agreed that the trainers provided the needed support all the time (Fig. 5).

The last part of the survey asked the students about their eagerness to learn more programming in an online context, and how they rate their programming level after the program completion. The results showed that 99% of the students are satisfied with their programming level after this program, while only 1% of the students reported that their programming levels are still weak. Still, when they asked whether they prefer to join the future programs online or on campus, 55% of students preferred to participate in online programs, 44% of students showed their desire to join the program on campus, while 2% of the students showed their unwillingness in participating any future program (Figs. 6 and 7).

Fig. 5 Trainers' support

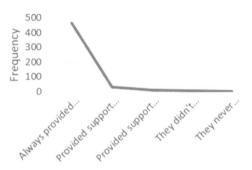

Did the trainers provide aid and support throughout the program on the various...

Fig. 6 Programming skills rate

'Frequency' by 'How would you rate your programming skills after completing this program?'

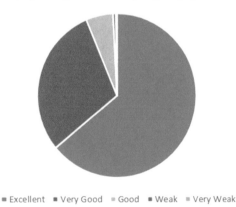

■ Excellent   ■ Very Good   ■ Good   ■ Weak   ■ Very Weak

Fig. 7 Eagerness to learn more programming

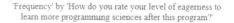

'Frequency' by 'How do you rate your level of eagerness to learn more programming sciences after this program?'

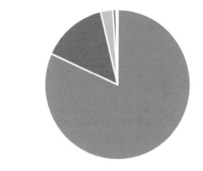

■ So Excited ■ Excited ▪ Neutral ■ Not Excited ▪ Not Excited at all

5   Discussion of The Results

This paper addresses the pedagogical concerns regarding learning coding in an online learning context; it aims to examine Emirati students' performance following a two-week online coding session. By observing the students' interaction during the program, the authors investigated the students' performance and challenges of online programming activities. The analysis suggests that most students had positive attitudes towards online coding and got many benefits from it; for instance, they can work at their own pace and in their own style to learn and experiment with the subject. Instructors, in addition, have less administrative work to do, and they can use the course content in various ways. However, despite the many benefits of this learning mode, some students experienced some form of technical difficulty in programming and a lack of interaction with the online environment.

Turning now to the survey results, two survey questions obtained a 100% agreement. As can be observed by looking at the results, these items are as follows: the students' programming skills have been advanced, and trainers have provided full support during the program. Additionally, the results showed 97% agreement of students' eagerness to join more programming courses in the future. These agreements show that the participants have positive perceptions towards learning to code and agreed that they got many benefits from the online learning environment. However, when they were asked whether they wanted to attend more programs online or in the traditional classroom, they separated into two sections; 54.6 showed a willingness to participate in the future courses online, while 43.6 preferred to participate in the traditional classes. Other results from the survey show that some barriers prevented a few students from fully understanding online coding. This can be supported by the results that show that these students are not willing to join more coding programs. By comparing these findings to the literature, it is clear that despite the rapid expansion of online learning environments and the numerous advantages they provide in terms of creating courses that simulate more traditional classroom settings, particularly in programming, more research is needed to better understand the difficulties and barriers students face when they learn to code online, and how can we overcome them.

6   Conclusion

This study used a mixed-method approach to determine how Emirati students perceived programming in an online learning environment. The analysis of the qualitative results revealed encouraging results on students' general perceptions of programming, with many achieving successes. While some reported technical challenges, the majority were more skilled than expected at resolving them. The findings indicated that students of this age could benefit from online learning programs. Also, the quantitative results validated the initial qualitative findings by elucidating all of the variables encountered by the Emirati students who participated in the program. The study suggests that a framework is needed to assist teachers in capturing student variety and implementing, evaluating, and determining what must be done in the online learning environment. Additionally, the authors recommend learning coding in various educational settings, such as the blended setting, where teachers are present and can facilitate the learning process.

This study added to the growing body of literature on the importance of programming skills for students and the impact of online learning environments in acquiring this skill, additionally, it paved the way for more research to integrate computational thinking and programming approaches in different learning approaches. The outcomes of this study may assist educational policymakers in better implementation of online learning, particularly in programming. The study can also help educators better prepare for the problems that students may confront with online programming and shed light on prospects for expansion within it, however, it has some limitations that may affect its generalizability. The participants included were all Emirati students having the same racial, cultural, and social backgrounds. Future studies are needed to conduct a similar investigation in different cultural contexts. The study also did not consider cheating and communication between students. It is necessary to determine whether online programming affects the quality of code presented by learners and the quality of communication with each other.

References

Abiodun, O. S., & Lekan, A. J. (2020). Children perceptions of the effectiveness of online coding as a supplement to in-person boot camps. *International Journal Scientific Advances, 1*(3), 187–191.

Benotti, L., Aloi, F., Bulgarelli, F. & Gomez, M. J. (2018). The effect of a web-based coding tool with automatic feedback on students' performance and perceptions. *SIGCSE 2018 - Proceedings of the 49th ACM Technical Symposium on Computer Science Education*, Baltimore, 21–24 February. Association for Computing Machinery, USA. https://doi.org/10.1145/3159450.315 9579. Accessed 29 Feb 2022

Boot, W. R. & Charness, N. (2016). Handbook of the Psychology of Aging. Schaie, K. W. & Willis, S. L. (eds). *The Academic Press*. San Diego. Ch.20.

Chao, P. Y. (2016). Exploring students' computational practice, design and performance of problem-solving through a visual programming environment. *Computers Education, 95*, 202–215.

Creswell, J. (2018). *Research Design: Qualitative, Quantitative & Mixed Methods Approaches* (5th ed.). Sage.

Davis, F. D. (1989). Perceived usefulness, perceived ease of use, and user acceptance of information technology. *MIS Quarterly, 13*(3), 319–340.

Daradoumis, T., MarquèsPuig, J. M., Arguedas, M., & CalvetLiñan, L. (2019). Analyzing students' perceptions to improve the design of an automated assessment tool in online distributed programming. *Computers Education, 128*, 159–170.

De, A. & Do, N. (2021). *Towards a Live Programming Platform for K-12*. Master. Dissertation. University of Porto

Galdo, A. C., Celepkolu, M., Lytle, N. & Boyer, K. E. (2022). Pair Programming in a Pandemic: Understanding Middle School Students' Remote Collaboration Experiences. *Mehmet Celepkolu.*

Kaplancali, U. T., & Chiu, Z. (2017). Teaching coding to children: a methodology for kids 5+. *International Journal Elementary Education, 6*(4), 32. https://doi.org/10.11648/j.ijeedu.201706 04.11

Kong, S. C., Chiu, M. M., & Lai, M. (2018). A study of primary school students' interest, collaboration attitude, and programming empowerment in computational thinking education. *Computers Education, 127*, 178–189.

Lewis, S. (2020). *Analysis of how primary-aged children learn to code: A Year 5 case study using Ev3 LEGO® robotics and stimulated recall*. Ph.D. Thesis. University of Central Queensland.

Manita, F., Durão, S. & Aguiar, A. (2021): Faculdade De Engenharia Da Universidade Do Porto Towards a Live Programming Platform for K-12.

Nariman, D. (2021). Impact of the interactive e-learning instructions on effectiveness of a programming course. *Advances in Intelligent Systems and Computing*. Springer, pp. 588–597

Ouahbi, I., Kaddari, F., Darhmaoui, H., Elachqar, A., & Lahmine, S. (2015). Learning basic programming concepts by creating games with scratch programming environment. *Procedia - Social Behavioral Sciences, 191*, 1479–1482.

Papert, S. (1972). Teaching children thinking. *Programmed Learning Educational Technology, 9*(5), 245–255.

Papert, S. (1980). Personal computing and its impact on education. *The computer in the school: Tutor, tool, tutee*, pp.197–202.

Stead, A. G., & Blackwell, A. (2014). Learning syntax as notational expertise when using Drawbridge. *Psychology of programming interest group annual conference*. University of Sussex. Brighton. 25–27 June.

Stephany, F., Braesemann, F. & Graham, M. (2021): Coding together–coding alone: the role of trust in collaborative programming. *Information Communication and Society.Routledge, 24*(13), 1944–1961.

Tsai, C. Y. (2019). Improving students' understanding of basic programming concepts through visual programming language: The role of self-efficacy. *Computers Human Behavior, 95*, 224–232.

Vico, F., Masa, J. & Garcia, R. (2019). ToolboX. Academy: Coding & Artificial Intelligence made easy for kids, Big Data for educators. *Proceedings of the 11th Annual International Conference on Education and New Learning Technologies*. Madrid, Spain.

Watson, C., & Li, F. W. (2014). Failure rates in introductory programming revisited. *Proceedings of the 2014 ACM conference on innovation & Technology in computer science education*, Uppsala. Sweden: ACM

A Quantitative Study on the Impact of Online Learning on Reading Comprehension Skills

Ranya Ahmed El Haddad and Sa'Ed Mohammad Issa Salhieh

Abstract This quantitative study aims to investigate the relationship between e-education and reading comprehension skills acquisition. It also examines if the previous relationship may impact students' results in the exams. It also analyses the relationship between students' knowledge in ICT and their perception and acceptance of online education. To collect data, A survey was sent to students to measure their perception of and satisfaction with online learning. Moreover, the marks of 105 students in an on-campus test were compared to the marks of another one they did online during the pandemic. The study agreed with the previous studies that e-learning can impact the reading skills positively and that students are getting aware of its educational benefits. On the other hand, the study did not agree with other studies about students' knowledge of ICT and how it can positively impact their perception of online education. The study showed that although secondary students have sufficient knowledge of ICT, they do not have positive perceptions of online education.

Purpose - to investigate the relationship between e-education and acquiring reading comprehension skills, and if this may impact students' results in the exams.

Methodology - A quantitative study in which a survey and the scores of two reading exams are analysed.

Findings - this study agreed with other studies about the positive impact of e-learning with some differences regarding students' satisfaction with IT.

Implications - teachers can integrate interactive websites within instruction and using online games and activities can make students more attentive and less distracted.

Originality/value - although most of the studies have proved that there is a positive relationship between the quality of ICT services and students' satisfaction with online education, this study disagrees as unlike most of the studies, the study in hand was conducted in a secondary school, not in a university.

Keywords Reading comprehension · Reading skills · Online reading · ICT

R. A. El Haddad (✉) · S. M. I. Salhieh (✉)
The British University in Dubai, Dubai, UAE
e-mail: 20000220@student.buid.ac.ae

S. M. I. Salhieh
e-mail: saed.salhieh@buid.ac.ae

© The Author(s) 2023
K. Al Marri et al. (eds.), *BUiD Doctoral Research Conference 2022*,
Lecture Notes in Civil Engineering 320,
https://doi.org/10.1007/978-3-031-27462-6_13

1 Introduction

During Covid19, most of the world has switched to distance education in fear of more spread of the pandemic. This sudden change has put most of school students in confusion which had an impact on their schooling attitude, hence their exam results. In the UAE, the government has tried their best to eliminate, or at least, lessen this fear by providing teachers with trainings required to overcome these non-precedential circumstances. They also provided schools with facilities and equipment to ensure students' easy accessibility of resources and materials needed to continue learning as smoothly as possible (Ati & Guessoum, 2010).

Reading comprehension is a complex skill taught online as a part of the English course delivered to secondary students. It requires connecting points to create a meaning or meanings that are partially derived from prior knowledge. It is an everyday skill that people practice all the time intentionally or unintentionally, yet at school, students should master reading comprehension skills that are developed in classrooms to understand all subjects and pass their exams (Destari, 2010).

1.1 Research Questions

1. Does on-line education have a significant relationship with students' levels in reading comprehension?

   A. Is there a significant difference between the scores of reading exams (on-campus and online)?
   B. Is there a relationship between the total scores of the two exams and the reading skills mastered in each learning situation?

2. Do students' level of Knowledge in ICT and the Benefits of online education have an impact on their Students' Rating of Online Education?

2 Literature Review

2.1 Conceptual Framework

Many concepts can be discussed in this section to give a comprehensive account of this topic such as: *reading comprehension* and *online education.* **Reading Comprehension** is the capability to read, process, and comprehend written material (Butterfuss et al., 2020). **Online Education** is the use of information technologies and communications to assist in the development and acquisition of knowledge from faraway areas (Basilaia & Kvavadze, 2020).

Fig. 1 The structural theory and reading comprehension

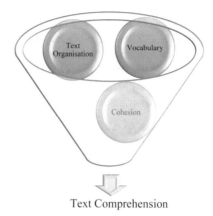

Text Comprehension

2.2 Theoretical Framework

Many theories have discussed reading comprehension, online learning and students' perception and attitudes. However, the study in hand will discuss the *Structural Theory*, *The Digital Native Theory*, and *Behaviourism.*

The Structural Theory: It is hard for L2 students to understand written texts for many reasons such as the limited vocabulary knowledge and the text structure and cohesion. Moreover, the text features can influence cognitive process that govern reading comprehension (Jake Follmer & Sperling, 2018) as shown in Fig. 1.

The Digital Native Theory: It is believed that nowadays students are digital natives as they were born during the digital revolution (Von der Heiden et al., 2011), so they prefer working and gaming online.

Behaviourism: A behaviour can be due to some external and/or internal causes (Whiteley, 1961). In this case, the external cause is the sudden shift to online education. On the other hand, Skinner (2011) identified Behaviourism as the philosophy of human behaviour. According to him, behaviour is not about cause-and-effect connection, but it is about a set of functional actions that take place in a certain order such as the pandemic, social distancing and e-learning.

2.3 Literature Review

The Benefits of Online Learning on Reading Comprehension: Recently, using technology has been proved successful in improving students' levels of performance in many subjects including reading comprehension. Many studies have been conducted to identify the type of impact of online education on reading comprehension. The studies of Zidat and Djoudi (2010) and Ciampa (2012), have proved that using technology, multimedia and games increased students' opportunities to gain more reading skills. Other studies have revealed the important role of online reading in improving the level of performance of poor readers which, consequently, improved their reading comprehension skills.

The Impact of Students' Satisfaction on Their Levels of Performance: Many studies have confirmed the positive relationship between students' satisfaction and behaviour, and their levels of performance in different contexts. In their studies, Sapri et al. (2009), Dhaqane and Afrah (2016) proved that teaching and learning methods used in the higher education institutes had a significant impact on students' satisfaction which consequently improved their levels of performance. Another study. Furthermore, the study conducted on Vietnamese College students, Salehi et al. (2014) found out that students with ICT knowledge can feel comfortable learning online.

3   Methodology

This quantitative study will examine the impact of online education on students' reading comprehension skills and the impact of their ICT knowledge on their satisfaction and behaviour towards online learning. To do so, the study will compare 10th graders' results in reading comprehension prior and during distance learning, and analyse the data collected via a survey that will be dispatched to the same students.

3.1   Quantitative Research Paradigms

Paradigms can be considered the 'worldview' or 'sets of beliefs' that govern the research approaches and methods and lead to answer the research questions (Cohen et al., 2018). It is suitable to discuss as it underpins the quantitative approach.

post-Positivism: This theory underpins the quantitative approach as it is concerned with numbers and statistics. According to Alakwe (2017), post-positivists believe that knowledge is extracted from data that is statistically analysed. This knowledge can be generalizable in similar contexts if showing the same reality observed. This theory is also concerned with decreasing human bias by testing pure data that is not yet interpreted by people.

3.2 Research Methods

There are two instruments used in this study: the first one is 105 10th graders' scores in 2 reading comprehension quizzes. The first one was administered at school before the pandemic and the second one was administered online during the pandemic to determine the significance in difference of means using descriptive data and 'Paired t-test' on SPSS.

The second tool was a survey to collect data from the same students regarding their attitudes toward the online education phenomenon, the challenges they might have faced while implementing the online education and the level of satisfaction. The survey was conducted anonymously to guarantee objectivity and privacy. The survey used Likert scale in all questions for easier collection of responses.

The survey was adopted from two published studies[1]: (Simpson, 2012; Al-Azawei & Lundqvist, 2015). Surveys are used to collect data in the quantitative approach due to the vast development in technology (Mathers et al., 2009).

3.3 Sampling

A sample is a part of the population chosen to represent the whole population. The population targeted is 10th graders, and the sample is 105 female students in a private school in Ajman. There are many types of sampling, but the researcher used the convenience sampling technique due to the nature and logistics of the study during the pandemic (Acharya et al., 2013).

4 Result Analysis and Discussion

The study showed that there is a significant positive relationship between online education and students' improvement in reading skill, yet their satisfaction with and perception of online education is not necessarily congruent with the ICT services provided.

---

[1] "Learner Differences in Perceived Satisfaction of an Online Learning: an Extension to the Technology Acceptance Model in an Arabic Sample" and "Student Perceptions of Quality and Satisfaction in Online Education".

4.1   Question No. 1 and Sub Questions A&B

To answer Q.1, sub-questions A&B will be answered first to be able to find out if there is an impact of online teaching on students' levels of performance in reading comprehension skills.

Sub-Question A: Is there a significant difference between the scores of reading exams (on-campus and online)? The null hypotheses are: H0: "there is no significant difference in mean between the scores of on-campus reading test and the online reading test" while the alternative hypothesis (H1) is: "there is a significant difference in mean between the scores of on-campus reading test and the online reading test'. A '**paired t-test**' was conducted to confirm or reject the null hypothesis (**H0**) (Table 1).

As the significance factor is P =.732 is higher than $\alpha = .05$ (P > α), it means that there is no statistically significant difference in means of the scores of the two tests, so the previous results failed statistically to reject the null hypothesis which states that "there is no significant difference in mean between the scores of on-campus' reading test and the online reading test" with 95% confidence.

Sub Question B: Is there a relationship between the total scores of the two exams and the reading skills mastered in each learning context? A correlation test will be used to answer the question.

To determine the relationship between the previous variables, correlation tests will be used. The null hypothesis (H0) is "there is no significant relationship between reading skills acquired in each educational context and the tests conducted". P = 0, while the alternative hypothesis (H1) is: "there is a significant relationship between reading skills acquired in each educational contexts and the tests conducted" P $\neq$ 0).

The following Tables 2 and 3, show an overall statistically significant positive relationship between the acquired reading skills and the scores of reading tests whether on-campus or online. There is also a significant difference in means between the reading skills acquired online and those acquired at school in favor for the online context.

Does on-line education have a significant relationship with students' levels in reading comprehension? The percentages of students' attendance will be used as a reflection of the impact of e-learning as students used to join classes every day. The hypotheses of this questions are the null hypothesis (H0) is: "There is no significant relationship between the percentage of students' attendance and their scores in the

Table 1   Paired t-test, the significance in difference...etc.

Paired samples correlations		N	Correlation	Significance	
				One-sided p	Two-sided p
Pair 1	On_campus_reading_score & Online_reading_scores	105	0.034	0.366	0.732

Table 2 The correlation between the skills acquired on-campus and the on-campus test scores

Correlations

		On_campus_reading_score	mastered_skills_T1
On_campus_reading_score	Pearson correlation	1	0.922**
	Sig. (2-tailed)		0.000
	N	105	105
mastered_skills_T1	Pearson correlation	0.922**	1
	Sig. (2-tailed)	0.000	
	N	105	105

**. Correlation is significant at the 0.01 level (2-tailed)

Table 3 The correlation between online acquired skills and online test scores

		mastered_skills_T3	Online_reading_scores
mastered_skills_T3	Pearson correlation	1	0.542**
	Sig. (2-tailed)		0.000
	N	105	105
Online_reading_scores	Pearson correlation	0.542**	1
	Sig. (2-tailed)	0.000	
	N	105	105

**. Correlation is significant at the 0.01 level (2-tailed)

online reading test". (p = 0), and the alternative hypothesis (H1) is: "There is a significant relationship between the percentage of students' attendance and their scores in the online reading test" (p ≠ 0). A Pearson correlation test was used to confirm or reject the null hypothesis (Table 4).

Coefficient (r) is 0.346. This shows a positive relationship, and it cannot be considered a relatively strong relationship as it is not close to +1. The p value is .001 < alpha value .05. This means that the results statistically reject the null hypothesis and confirms the alternative hypothesis (H1): "Statistically, there is a significant relationship between the percentage of students' attendance and their scores in the online reading test" Consequently, all the previous results of question 1 and the sub questions A&B prove the fact that there is an overall positive significant relationship between online education and reading comprehension skills acquired and the overall online reading tests score. The previous results conform with Zidat and Djoudi (2010) and Ciampa (2012) that the online education is beneficial in relation to reading comprehension skills acquisition.

Table 4 The correlation between students' percentage of attendance and online reading test scores

Correlations

		attendance_percentage	Online_reading_scores
attendance_percentage	Pearson correlation	1	0.346[**]
	Sig. (2-tailed)		< 0.001
	N	105	105
Online_reading_scores	Pearson correlation	0.346[**]	1
	Sig. (2-tailed)	< 0.001	
	N	105	105

**. Correlation is significant at the 0.01 level (2-tailed)

4.2   Do Students' Level of Knowledge in ICT and the Benefits of Online Education Have an Impact on Students' Rating of Online Education?

A survey was conducted on n = (105) to measure students' knowledge in and satisfaction with ICT. The null hypothesis (H0) is: "There is no significant correlation between students' level of knowledge in ICT and their Recognition of the online education benefits on their overall rating of online education". The alternative hypothesis (H1) is: "There is a significant correlation between students' level of knowledge in ICT and their Recognition of the online education benefits on their overall rating of online education."

A Linear Regression test was conducted to get answers to the previous question (Table 5).

The previous table shows that: P value of the predictor ICT is .432 > alpha value .05 which means that the relationship between ICT and students' satisfaction is not significant, yet the relationship between the Benefits of online education and satisfaction is significant as P =.001 < α = .05, so there is a significant relationship

Table 5 The relationship between students' ICT knowledge and their satisfaction with online reading

Coefficients[a]

Model		Unstandardized coefficients		Standardized coefficients	t	Sig.	95.0% confidence Interval for B	
		B	Std. error	Beta			Lower bound	Upper bound
1	(Constant)	5.550	1.179		4.706	<0.001	3.211	7.889
	Benefits	0.208	0.039	0.466	5.334	<0.001	0.131	0.286
	ICT_Logistics	−0.063	0.080	−0.069	−0.788	0.432	−0.221	0.095

[a]Dependent Variable: Satisfaction

between the benefits of online education and students' satisfaction which conforms with Whiteley (1961) that their satisfaction (effect) is a result of the benefits they are aware of (cause), yet there is no significant relationship between ICT knowledge and students' satisfaction. This agrees with Skinner (2011) as students' negative behaviour and perception of online education is not a result of their lack of knowledge, and it can be a philosophy that has emerged due to other emotional and social factors such as the lack of socialization caused by distance learning.

5 Conclusion, Implications, and Research Significance

Conducting the previous tests, some findings can be highlighted, and some conclusions can be made accordingly.

5.1 *Conclusion*

The tests conducted show that:

- There is a significant impact of online learning on improving the reading comprehension tests' scores.
- These results have confirmed that there is a significant impact of reading skills gained in both educational contexts and the reading tests scores in both contexts.
- The study also has proven that there is a positive relation between students' satisfaction with online education and their improvement in reading skills, yet the relationship between students' perceptions of online education and the ICT services provided to them.

5.2 *Implications and Suggestions*

Implications: The hypotheses confirmed in this study can indicate that the types of teaching materials can have a great impact on students' satisfaction and performance. Using versatile activities and different websites can decrease the boredom and monotony that students might feel in actual classrooms.

Suggestions: Using creative reading material can motivate students to study and practice, so it will be much better to use online reading comprehension resources and activities even after going back to school. Moreover, students can have the chance to study in virtual classes and practice e-reading activities even when they are back to school for at least one school class. This will enable students to enjoy reading and practicing using reading comprehension skills more effectively.

5.3 Research Significance

The results of the study agree partially with the previous studies in that domain, yet it does not agree with the results of other studies about the impact of ICT services on students' satisfaction with online education. As most studies focused on tertiary students. This study can encourage other researchers to further investigate the context of high school students' satisfaction and its relationship with ICT services which might reveal new dimensions that might enrich research and become new references to other scholars.

References

Acharya, A., Prakash, A., Saxena, P., & Nigam, A. (2013). Sampling: why and how of it? *Indian Journal of Medical Specialities, 4*(2), 330–333.

Alakwe, K.O. (2017). Positivism and knowledge inquiry: from scientific method to media and communication research. *Specialty Journal of Humanities and Cultural Science, 2*(3), 38–46. Accessed May 31, 2021.

Al-Azawei, A. & Lundqvist, K. (2015) Learner differences in perceived satisfaction of an online learning: an extension to the technology acceptance model in an Arabic sample. *Electronic Journal of e-Learning, 13*(5), 412–430. Accessed June 6, 2021.

Prensky, M. (2001). Digital natives, digital immigrants part 1. *On the Horizon, 9*(5), 1–6.

Aliaga, M. & Gunderson, B. (2003). *Interactive statistics* (2nd ed.). Pearson education, Inc.

Ati, M. & Guessoum, N. (2010). E-learning in the United Arab Emirates. https://www.researchg ate.net/publication/233428047_E-Learning_in_the_United_Arab_Emirates. Accessed May 25, 2021.

Boudalia, M. (2018). Effect of early exposure to technology on student satisfaction with online education. Ph.D. in Management and Technology. Walden University.

Butterfuss, R., Kim, J. & Kendeou, P. (2020). Reading comprehension. Oxford Research Encyclopedia of Education. Accessed May 28, 2021.

Cavanaugh, C., Barbour, M. & Clark, T. (2009). Research and practice in K-12 online learning: a review of open access literature. *The International Review of Research in Open and Distributed Learning, 10*(1).

Basilaia, G. & Kvavadze, D. (2020). Transition to online education in schools during a SARS-CoV-2 coronavirus (COVID-19) pandemic in Georgia. *Pedagogical Research, 5*(4). Accessed June 20, 2022.

Ciampa, K. (2012). ICANREAD. *Journal of Research on Technology in Education, 45*(1), 27–59.

Cohen, L., Manion, L., & Morrison, K. (2018). *Research methods in education*. Routledge.

Coiro, J. (2011). Predicting reading comprehension on the internet. *Journal of Literacy Research, 43*(4), 352–392.

D'Elia, E. (2005). Using the results of qualitative surveys in quantitative analysis. *SSRN Electronic Journal*.

Destari, D. (2010). The effectiveness of internet-based material to teach reading comprehension viewed from learning motivation. graduate degree. Sebelas Maret University Surakarta.

Dhaqane, M. & Afrah, N. (2016). Satisfaction of students and academic performance in Benadir university. *Journal of Education and Practice, 7*(24), 59–63. Accessed June 5, 2021.

Evans, J., & Mathur, A. (2005). The value of online surveys. *Internet Research, 15*(2), 195–219.

Fidalgo, P., Thormann, J., Kulyk, O. & Lencastre, J. (2020). Students' perceptions on distance education: a multinational study. *International Journal of Educational Technology in Higher Education, 17*(1), 1–18. Accessed June 13, 2021.

Jake Follmer, D. & Sperling, R. (2018). Interactions between reader and text: contributions of cognitive processes, strategy use, and text cohesion to comprehension of expository science text. *Learning and Individual Differences, 67*, 177–187. Accessed November 24, 2021.

Keegan, D. (1996). Foundations of distance education.

Mathers, N., Fox, N. & Hunn, A. (2009). Surveys and questionnaires. NIHR RDS for the East Midlands/Yorkshire & Humber. Accessed June 14, 2021.

Rezaee, A. & Sharbaf Shoar, N. (2011). Investigating the effect of using multiple sensory modes of glossing vocabulary items in a reading text with multimedia annotations. *English Language Teaching, 4*(2), 25. Accessed June 4, 2021.

Ridzuan, A., Yunus, M., Abdullah, M., Bakar, M. & Ramlan, A. (2018). The relationship between students satisfaction and their academic performance among public relations degree students in UiTM Alor Gajah Melaka. *International Journal of Academic Research in Business and Social Sciences, 8*(10), 874–882. Accessed June 5, 2021.

Salehi, H., Shojaee, M. & Sattar, S. (2014). Using E-learning and ICT courses in educational environment: a review. *English Language Teaching, 8*(1), 63–70. Accessed June 13, 2021.

Sapri, M., Kaka, A. & Finch, E. (2009). Factors that influence student's level of satisfaction with regards to higher educational facilities services. *Malaysian Journal of Real Estate, 4*(1), 34–51. Accessed June 5, 2021.

Simpson, J. (2012). Student perceptions of quality and satisfaction in online education. Doctor of Philosophy. University of Alabama.

Skinner, B. (2011). About behaviorism. New York: Random House

Sobh, R., & Perry, C. (2006). Research design and data analysis in realism research. *European Journal of marketing, 40*(11/12), 1194–1209.

Sterling, K. (2015). Student satisfaction with online learning. Ph.D. in Philosophy. University Of California, Santa Barbara.

Surahman, E. & Sulthoni. (2020). Student satisfaction toward quality of online learning in Indonesian higher education during the Covid-19 pandemic. In *6th international conference on education and technology*. Accessed May 28, 2021. https://ieeexplore.ieee.org/abstract/document/9276630

van den Broek, P., Rapp, D., & Kendeou, P. (2005). Integrating memory-based and constructionist processes in accounts of reading comprehension. *Discourse Processes, 39*(2–3), 299–316.

Von der Heiden, B., Fleischer, S., Richert, A., & Jeschke, S. (2011). Theory of digital natives in the light of current and future E-learning concepts. *International Journal of Emerging Technologies in Learning (IJET), 6*(2), 37.

Whiteley, C. (1961). II—behaviourism. *Mind, LXX*(278), 164–174.

Zidat, S., & Djoudi, M. (2010). Effects of an online learning on EFL university students' English reading comprehension. *International Review on Computers and Software, 5*(2), 186–192.

Simulation Study on the Effect of Courtyards Design on Natural Ventilation: The Case Study of a Beauty Centre in Germany

Bana Eid and Hanan M. Taleb

Abstract This paper examines how to enhance the indoor environmental quality and thermal comfort in a beauty centre in Germany, by implementing courtyard in the building as a natural ventilation method. The courtyard ventilation method will be discussed in the literature review and will apply it using IES VE computer software with different strategies in different scenarios to achieve the optimum natural ventilation with the best scenario. This will include designing the base case scenario, and next, the HVAC system will be switched off during the summer and apply 3 different strategies with 3 scenarios each to study the effect of natural ventilation of courtyards in terms of energy consumption and thermal comfort and choose the best-case scenario. The winning case scenario is "scenario 8" as it has dropped 2.3 MWh from the total energy based on the base case scenario. Natural ventilation is important as it helps in improving the thermal comfort of the indoor and outdoor environment and controlling the inside temperature along with the air quality and movement. Moreover, it helps in reducing the usage of energy and thus reducing costs. All the strategies, scenarios and simulations were done by the authors using IES VE software.

Keywords Courtyards · Natural ventilation · Thermal comfort · Air quality · Solar gain

1 Introduction

Nowadays, the awareness of excessive use of energy and its cost and environmental impact have been increased. However, achitects, engineers, and designers prefer providing building designs that suit the climatic condition and adapt to nature.

B. Eid (✉)
Architecture and Sustainable Built Environment, The British University in Dubai, Dubai, United Arab Emirates
e-mail: bana2610@hotmail.com; 20000005@student.buid.ac.ae

H. M. Taleb
Faculty of Engineering, The British University in Dubai, Dubai, United Arab Emirates

© The Author(s) 2023
K. Al Marri et al. (eds.), *BUiD Doctoral Research Conference 2022*,
Lecture Notes in Civil Engineering 320,
https://doi.org/10.1007/978-3-031-27462-6_15

Applying natural ventilation methods in buildings that suit the climate and environment is important to provide the cooling and heating required with environmentally friendly passive design features and less energy consumption.

Natural ventilation has different methods depends on the building's design, location, and climatic conditions, such as, single- sided ventilation, cross ventilation, stack ventilation and courtyard ventilation. In this study the focus will be on the courtyard method and its effect on the building and energy use. The courtyards method has been recognized as an option to provide healthier and comfortable environment with minimal energy consumption.

This paper introduces the natural ventilation systems along with the different methods used to achieve it. The courtyards method will be discussed in detail to provide natural ventilation in the buildings and achieve the indoor environmental quality and thermal comfort. A case study of beauty centre building in Germany will be introduced to study the energy consumption of the building when the HVAC system is working all over the year. Following to this, a courtyard method will be implemented in the building in 3 different scenarios to study the natural ventilation system in it and compare the results of the energy consumption outcomes with and without courtyard. This study and analysis will be done using IES VE computer software.

2   Literature Review

Natural ventilation can be defined as a method of driving outdoor fresh air to the indoor environment of a building through building's envelop using passive forces. The main driving natural forces that affect natural ventilation in buildings are thermal buoyancy and wind (Ventive n.d.). Courtyards are one of the methods that can be used as a passive design to achieve natural ventilation (Krarti, 2018).

2.1   Definition of Courtyard

Reference to Oxford's Dictionary, it defines courtyards as an unroofed space that is surrounded entirely or partially by walls, houses or buildings (Oxford University n.d.). On the other hand, the definition of a courtyard in the Cambridge Dictionary is a flat ground outside place that is entirely or partially surrounded by a building's walls (Cambridge University n.d.). Both definitions agree that the courtyards are an open area surrounded by buildings or wall and has no coverage.

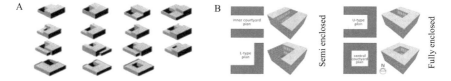

Fig. 1 A: Different applied design forms for courtyard of one or two storey building, B: Fully and semi enclosed design of courtyards (Adapted from (Abass, Ismail & Solla, 2016))

2.2 Courtyard Forms and Elements

Courtyards are flexible with their design, they can be designed in any shape, but the usual design is square, rectangular, and circle. These forms have been designed to achieve environmental aspects like topography, site limitation, building orientation and function to provide different forms such as [U shape, L shape, T shape, V shape, H shape or Y shape]. Different design forms can be applied for the courtyard of one or two storey building as shown in Fig. 1. The main purposes of designing a courtyard are to provide natural light, fresh air through natural ventilation, play area for the children, and for rest and activities (Abass, Ismail & Solla, 2016).

2.3 Characteristics of Courtyard in Four Climates

The characteristics of courtyard design can vary in different climate conditions in order to function as a passive design strategy for optimum use of natural element such as daylight and wind. Table 1 shows the different characteristics of courtyard design in different climate condition. The courtyards are used as a source of daylighting, natural ventilation, and natural heating and cooling (Taleghani, Tenpierik & van den Dobbelsteen, 2012).

3 Research Methodology

This research will start by providing a literature review about the natural ventilation system including definition, different types of natural ventilation and its impact on thermal comfort and energy consumption. The courtyard method as one of the natural ventilation types will be selected to explore more and discuss it in detail. Moreover, the courtyard method will be applied on a case study of beauty centre building in Germany to analyze and compare the natural ventilation effect in the building with and without the implementation different types of courtyards in 3 scenarios. IES VE software will be used for the computer simulation, analysis, and results.

Table 1 The different characteristics of courtyard design in different climate condition

Characteristics	Hot arid climate	Cold climate	Temperate climate	Tropical climate
Building design shape	- Introverted - Southern block of the building is the biggest - The highest ratio of void to solid	- Introverted - Northern block of the building is the biggest - The lowest ratio of void to solid	Small, narrow, and deep courtyards for easy stack effect	- Extroverted - Size of blocks are equal - Hight of building is high
Natural elements within the courtyard	- Trees - Water element (fountain) - Shrub and lawn	N/A	Limited elements	Limited trees
Openings in walls surrounding the courtyard and facades	Small size vertical windows	Small and limited openings	Large openings	Large openings

Source (Taleghani, Tenpierik & van den Dobbelsteen, 2012)

3.1 Climate Analysis

Germany has variation in the climate analysis due to various mountains and hill areas over the country. However, the climate in Berlin characterized by a maritime temperate climate. In the summer, Berlin's temperature gets high along with many warm sunshine hours between May and August. July is the peak of the summer in Berlin where the temperature reaches 25 C°, while the lowest amount of sunlight is in December. The temperatures range between 20 C° to 25 C°. July is the warmest month with average maximum temperature of 25 C°. While in the winter, the weather gets cold with maximum temperature of 3 C° in January (World Weather & Climate, 2021). For the wind, the month of January has the most wind average, and August has the least wind average. The most prevailing wind is from the south-west direction, and the annual average speed is 9.6 mph (IEM: Site Wind Roses, 2020).

3.2 Case Study and Project Detail

The case study is about a beauty centre building located in Berlin, Germany. The building consists of ground floor plan with one main entrance, operable windows on the external walls and different room functions inside it as shown in Fig. 2. The total floor area is 110 m².

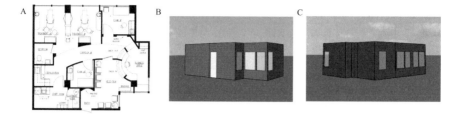

Fig. 2 A: Ground floor plan of beauty centre building in Berlin, Germany (*Source* (Taleb 2021, via IES software)), B: Front view of the beauty centre, C: Rear view of the beauty centre, Source (Eid 2021, via IES software)

3.3 Modelling using IES VE software

Building Code and Specification Standards

Berlin has specific building energy efficiency standards and policies that are required in designing and constructing new buildings. The supporting policies covers different building aspects for the U-values of roof, wall, floor, and window. The U-values required for new buildings are: roof is 0.2 W/m^2K, wall and floor is 0.28 W/m^2K, and window is 1.3 W/m^2K (Energy Conservation Regulations (EnEV) I Global Buildings Performance Network n.d.). The model set-up via IES VE software includes: setting out location and weather data, creating construction material for walls, floors, roof and windows, creating profile for cooling, heating and occupancy, setting out thermal template - Apache system, internal gains, and air exchanges.

The Base Case Model Design via IES VE Software

Figure 2 shows the ground floor plan and the three-dimensional base case model of the beauty centre building using the mentioned U-values.

The Optimal Orientation

Selecting the optimal orientation was done through Sun cast simulation in the IES software. Table 2 shows the total energy consumption of each direction. Reference to Apache simulation results, the optimal orientation is 180 degrees from the north (south direction) is the optimal orientation as it has the lowest energy consumption.

3.4 Design Strategies

The focus will be on three design strategies for courtyards to study their effect on natural ventilation. The first strategy is the courtyard location, second is the percentage of openable windows area around the courtyard, and third is Courtyard proportions from x to 2x as width, where x is the courtyard height. The simulation was done on July 14 for all strategies as it is the hottest day.

Table 2 Total energy consumption of the base case in different directions

Date	Total energy (MWh)	Total energy (MWh)	Total energy (MWh)	Total energy (MWh)	Total energy (MWh)	Total energy (MWh)	Total energy (MWh)	Total energy (MWh)
	North orientation	45 from the north	90 from the north	135 from the north	180 from the north	225 from the north	270 from the north	315 from the north
Jan 01–31	2.6182	2.61	2.6125	2.5898	2.5711	2.5987	2.6385	2.6385
Feb 01–28	2.2945	2.2815	2.2748	2.2416	2.2158	2.2615	2.3212	2.3262
Mar 01–31	1.5259	1.4984	1.4567	1.4113	1.3952	1.4628	1.5423	1.566
Apr 01–30	0.9202	0.9053	0.8672	0.8572	0.8914	0.9219	0.9269	0.9254
May 01–31	0.6159	0.6215	0.6292	0.6305	0.633	0.6427	0.635	0.6197
Jun 01–30	0.6485	0.6673	0.6962	0.6968	0.6747	0.681	0.6863	0.6666
Jul 01–31	0.7378	0.7597	0.7935	0.7945	0.7655	0.7647	0.7686	0.7526
Aug 01–31	0.4828	0.4889	0.5005	0.5032	0.4958	0.494	0.4929	0.4864
Sep 01–30	0.418	0.4166	0.4145	0.4137	0.414	0.4167	0.4196	0.4209
Oct 01–31	0.9272	0.9139	0.8988	0.8724	0.8541	0.8944	0.9485	0.956
Nov 01–30	1.8917	1.8747	1.8776	1.8374	1.8	1.849	1.9259	1.9303
Dec 01–31	2.3691	2.3629	2.3658	2.3517	2.3371	2.3536	2.3821	2.3829
Summed total	15.4497	15.4006	15.3873	15.2	**15.0477**	15.3409	15.6878	15.6715

Source (Eid 2021, via IES software)

Strategy 1: Courtyard Location

Summary of Strategy 1: Courtyard Location

Reference to Fig. 3 and Tables 3, 4 and 5, the best-case scenario in strategy one is scenario 1, as it is the lowest in total energy consumption, and it is better in terms of air temperature, relative humidity, solar gain and CO_2 concentration.

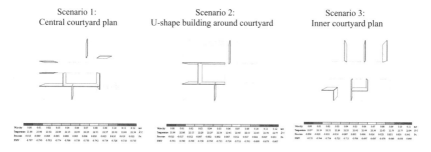

Fig. 3 Layout plan of scenario 1, 2 and 3, *Source* (Eid 2021, via IES software)

Table 3 Summary of scenario 1

Var. Name	Type	Min. Val	Min. Time	Max. Val	Max. Time	Mean
Air temperature	Temperature (C)	8.76	01:30,17/Jun	33.4	17:30,05/Jun	18.78
Relative humidity	Percentage (%)	24.4	18:30,05/Jun	99.98	01:30,02/Sep	67.12
Solar gain	Gain (kW)	0	00:30,01/Jan	3.3669	12:30,06/Mar	0.459
Room CO2 concentration	CO2 concentration (ppm)	400	02:30,01/Sep	1703	00:30,01/Jan	1377

Scenario 1 - Total Energy (MWh): 11.3364

Source (Eid 2021, via IES software)

Table 4 Summary of scenario 2

Var. Name	Type	Min. Val	Min. Time	Max. Val	Max. Time	Mean
Air temperature	Temperature (C)	8.55	02:30,17/Jun	33.49	17:30,05/Jun	18.53
Relative humidity	Percentage (%)	24.21	18:30,05/Jun	99.98	01:30,02/Sep	70.21
Solar gain	Gain (kW)	0	00:30,01/Jan	4.2518	12:30,06/Mar	0.5656
Room CO2 concentration	CO2 concentration (ppm)	400	03:30,01/Jun	1802	10:30,25/May	1450

Scenario 2 - Total Energy (MWh): 11.9551

Source (Eid 2021, via IES software)

Table 5 Summary of scenario 3

Var. Name	Type	Min. Val	Min. Time	Max. Val	Max. Time	Mean
Air temperature	Temperature (C)	8.65	01:30,17/Jun	33.62	17:30,05/Jun	18.72
Relative humidity	Percentage (%)	24.13	18:30,05/Jun	99.99	01:30,02/Sep	71.38
Solar gain	Gain (kW)	0	00:30,01/Jan	5.1305	12:30,06/Mar	0.6182
Room CO2 concentration	CO2 concentration (ppm)	400	02:30,01/Jun	1912	09:30,04/Feb	1533
Scenario 3 - Total Energy (MWh): 12.6353						

Source (Eid 2021, via IES software)

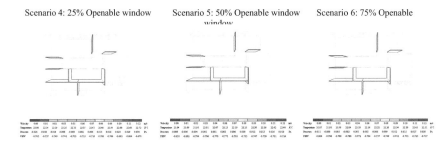

Fig. 4 Layout plan of scenario 4, 5 and 6, *Source* (Eid 2021, via IES software)

Strategy 2: Percentage of the Openable Window Area Around the Courtyard

Summary of Strategy 2: Percentage of the openable window area around the courtyard

Reference to Fig. 4 and Tables 6, 7 and 8, the best-case scenario in strategy two is scenario 4, as it is the lowest in total energy consumption, and it is better in terms of air temperature, relative humidity, solar gain, and CO_2 concentration.

Strategy 3: Courtyard proportions from x to 2x as width, where x is the courtyard height

Summary of Strategy 3: Courtyard proportions from x to 3x as width, where x is the courtyard height

Reference to Fig. 5 and Tables 9, 10 and 11, the best-case scenario in strategy three is scenario 8, as it is the lowest in total energy consumption, and it is better in terms of air temperature and solar gain.

Table 6 Summary of scenario 4

Var. Name	Type	Min. Val	Min. Time	Max. Val	Max. Time	Mean
Air temperature	Temperature (C)	8.72	09:30,10/Jan	33.95	17:30,05/Jun	18.91
Relative humidity	Percentage (%)	24.04	18:30,05/Jun	99.98	01:30,02/Sep	66.97
Solar gain	Gain (kW)	0	00:30,01/Jan	4.2518	12:30,06/Mar	0.406
Room CO_2 concentration	CO_2 concentration (ppm)	400	03:30,01/Jun	1802	10:30,25/May	1360

Scenario 4 - Total Energy (MWh): 12.2363

Source (Eid 2021, via IES software)

Table 7 Summary of scenario 5

Var. Name	Type	Min. Val	Min. Time	Max. Val	Max. Time	Mean
Air temperature	Temperature (C)	8.41	02:30,17/Jun	33.29	17:30,05/Jun	18.5
Relative humidity	Percentage (%)	24.3	18:30,05/Jun	99.98	01:30,02/Sep	70.3
Solar gain	Gain (kW)	0	00:30,01/Jan	4.2518	12:30,06/Mar	0.5656
Room CO_2 concentration	CO_2 concentration (ppm)	400	03:30,02/Jun	1802	10:30,25/May	1449

Scenario 5 - Total Energy (MWh): 12.465

Source (Eid 2021, via IES software)

Table 8 Summary of scenario 6

Var. Name	Type	Min. Val	Min. Time	Max. Val	Max. Time	Mean
Air temperature	Temperature (C)	8.48	02:30,17/Jun	33.39	17:30,05/Jun	18.52
Relative humidity	Percentage (%)	24.25	18:30,05/Jun	99.98	01:30,02/Sep	70.25
Solar gain	Gain (kW)	0	00:30,01/Jan	4.2518	12:30,06/Mar	0.5656
Room CO_2 concentration	CO_2 concentration (ppm)	400	04:30,01/Jun	1802	10:30,25/May	1449

Scenario 6 - Total Energy (MWh): 12.6351

Source (Eid 2021, via IES software)

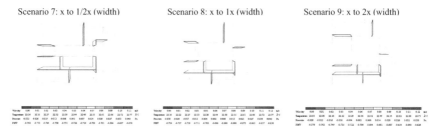

Fig. 5 Layout plan of scenario 7, 8, and 9, *Source* (Eid 2021, via IES software)

Table 9 Summary of scenario 7

Var. Name	Type	Min. Val	Min. Time	Max. Val	Max. Time	Mean
Air temperature	Temperature (C)	8.9	09:30,10/Jan	33.86	17:30,05/Jun	18.51
Relative humidity	Percentage (%)	24.11	18:30,05/Jun	99.98	02:30,02/Sep	68.15
Solar gain	Gain (kW)	0	00:30,01/Jan	4.7096	12:30,06/Mar	0.7692
Room CO2 concentration	CO2 concentration (ppm)	400	08:30,01/Jun	1681	00:30,01/Jan	1362
Scenario 7 - Total Energy (MWh): 14.2714						

Source (Eid 2021, via IES software)

Table 10 Summary of scenario 8

Var. Name	Type	Min. Val	Min. Time	Max. Val	Max. Time	Mean
Air temperature	Temperature (C)	8.42	09:30,10/Jan	33.97	17:30,05/Jun	18.6
Relative humidity	Percentage (%)	23.84	18:30,05/Jun	99.98	02:30,02/Sep	67.66
Solar gain	Gain (kW)	0	00:30,01/Jan	4.9488	12:30,06/Mar	0.5949
Room CO2 concentration	CO2 concentration (ppm)	400	02:30,01/Jun	1681	00:30,01/Jan	1361
Scenario 8 - Total Energy (MWh): 12.7703						

Source (Eid 2021, via IES software)

3.5   IES VE Simulation Results

To summarize, it was clear that the predicted mean vote, external velocity, and external pressure almost did not change through the three strategies. The winning case scenario has total energy (MWh): 12.7, while the base case scenario has 15.0477

Table 11 Summary of scenario 9

Var. Name	Type	Min. Val	Min. Time	Max. Val	Max. Time	Mean
Air temperature	Temperature (C)	8.9	02:30,17/Jun	33.59	17:30,05/Jun	18.4
Relative humidity	Percentage (%)	24.1	18:30,05/Jun	99.97	08:30,15/Oct	60.15
Solar gain	Gain (kW)	0	00:30,01/Jan	5.2373	12:30,06/Mar	0.9237
Room CO_2 concentration	CO_2 concentration (ppm)	400	02:30,01/Jun	1256	10:30,05/Jan	1043

Scenario 9 - Total Energy (MWh): 20.9776

Source (Eid 2021, via IES software)

MWh. For the air temperature, there was a slight change, but however, the temperature can be dropped in Berlin during the summer, and there will be a need for heating elements. In cold climates the need of ventilation is less, that is why the size of courtyards for one or two storey buildings are smaller in order to reduce the heat losses caused by it. In contrast, during winter season the natural heating is required for the building along with natural day light all the year.

4   Conclusion

Courtyards as a natural ventilation method are very useful as they allow for air circulation to enhance the indoor air quality. In addition, they provide daylight as they do not have top roof. Ventilated courtyards can be designed and applied for wide buildings whether they are residential or commercial buildings. Different cooling elements can be added to the courtyard to provide and enhance the natural ventilation, indoor temperature and solar gain.

This paper has been used IES software for investigating the performance of courtyards as a natural ventilation method within a Beauty centre in Berlin, Germany. Different strategies and scenarios have been implemented to reach the optimum natural ventilation performances. The design factors include shape and location of courtyard, the percentage of window openings around the courtyard, and courtyard proportions from x to 2x as width, where x is the courtyard height. To conclude, implementing courtyards in the buildings will enhance the indoor environment like the air quality and daylight, and thus improve the building's occupants' thermal comfort. However, different factors should be considered prior implementing courtyard design in any building. Building type and location along with the climatic conditions have major effect on the courtyards design and function. Different climatic conditions require different characteristics of courtyards design for optimum function. All these factors should be studied and considered in the design phase of a building.

References

Abass, F., Ismail, L. & Solla, M. (2016). Review of Courtyard House in Nigeria: definitions, history, evolution, typology, and functions. *AFRREV STECH: An International Journal of Science and Technology, 5*(2), 103

Dictionary.cambridge.org. (2021). *courtyard.* https://dictionary.cambridge.org/dictionary/english/courtyard

Global Buildings Performance Network. (2021). *GBPN—Energy Conservation Regulations (EnEV).* https://www.gbpn.org/databases-tools/bc-detail-pages/germany

IEM: Site Wind Roses. (2020). https://mesonet.agron.iastate.edu/sites/windrose.phtml?station=EDLW&network=DE__ASOS. Accessed January 28, 2021

Krarti, M. (2018). *Natural Ventilation—an overview | ScienceDirect Topics.* Sciencedirect.com. https://www.sciencedirect.com/topics/engineering/natural-ventilation

Oxford University. (n.d.). *courtyard noun—Definition, pictures, pronunciation and usage notes | Oxford Advanced Learner's Dictionary at OxfordLearnersDictionaries.com.* https://www.oxfordlearnersdictionaries.com/definition/english/courtyard

Taleghani, M., Tenpierik, M. & van den Dobbelsteen, A. (2012), Environmental impact of courtyards—a review and comparison of residential courtyard buildings in different climates. *Journal of Green Building, 7*(2), 113–136. https://meridian.allenpress.com/jgb/article/7/2/113/116408/ENVIRONMENTAL-IMPACT-OF-COURTYARDS-A-REVIEW-AND

Ventive. (n.d.). *What is Natural Ventilation? | Ventive.* https://ventive.co.uk/resources/natural-ventilation/

World Weather and Climate. (2021). *Climate in Berlin (Berlin Federal State), Germany.* https://weather-and-climate.com/average-monthly-Rainfall-Temperature-Sunshine,Berlin,Germany

Walker, A. (2016). Natural Ventilation | WBDG - Whole Building Design Guide. Wbdg.org. https://www.wbdg.org/resources/natural-ventilation

A Sustainable Approach to Improve the Interior Design of Existing Space: The Case of the BUiD Main Lobby

Rahaf Aloudeh, Manar Elmardi, and Wael Sheta

Abstract The Interior designers' role in the built environment is creating spaces that meet the needs of their clients through a process of critical thinking, research, and creative solutions. In addition to considering other aspects such as the client's health, safety, and wellbeing as well as the environmental impact which is considered a sustainable thinking by the designer. This paper will focus on analysing the interior characteristics for a space in an educational building. The study will thoroughly explore the existing condition of the chosen space, identify and analyse the raw materials, manufacturing methods, transportation, use, maintenance, and disposal of all the furnishings, fixtures, and equipment specified for an interior environment. Subsequently, the proposed strategies and methods will be presented to replace the existing conditions with sustainable alternatives. In addition to propose options for an alternative design that follows more sustainably responsible practices using the life cycle approach, explaining the solutions that could be used to improve sustainability of the chosen building interior space.

Keywords Sustainable interior design · Indoor air quality · ArchiCAD software · DesignBuilder · LCA Tools

1 Introduction

A sustainable interior design is a design that aligns with the main sustainable design principles and strategies that aims to design a built interior environment that provides the occupants with utmost comfort, meets the safety and health requirements, and achieve the economic savings and environmental responsibility (Hayles 2015). With the huge gap between principles and the reality of practices, it has become essential

R. Aloudeh (✉) · M. Elmardi · W. Sheta
Faculty of Engineering and IT, The British University in Dubai, Dubai, UAE
e-mail: 21002646@student.buid.ac.ae

W. Sheta
e-mail: wael.sheta@buid.ac.ae

© The Author(s) 2023
K. Al Marri et al. (eds.), *BUiD Doctoral Research Conference 2022*,
Lecture Notes in Civil Engineering 320,
https://doi.org/10.1007/978-3-031-27462-6_16

to connect the designers and their designs to the core meaning of sustainable interior design; in order to achieve the social, economic, and environmental goals of the design, and enhance the overall performance of the building in terms of energy-efficiency and interaction with surrounding environment over the lifecycle of the materials. The Interior designers' role in the built environment is creating spaces that meet the needs of their clients through a process of critical thinking, research, and creative solutions. In addition to considering other aspects such as the client's health, safety, and wellbeing as well as the environmental impact which is considered a sustainable thinking by the designer. The Life Cycle Assessment (LCA) is the method adopted by designers to achieve a longer, healthier and a considerate interior environment for the occupants (Linhares and Pereira 2017). This paper will focus on analysing the interior characteristics for a space in an educational building and proposing options for an alternative design that follows more sustainably responsible practices. The study will thoroughly explore the existing condition of the chosen space which includes layout, finishing materials, lighting, ventilation, furnishing and decoration elements. Subsequently, the proposed strategies and methods will be presented to replace the existing conditions with sustainable alternatives.

1.1 Aims and Objectives

The main aim of this paper is to create an overall environmentally friendly, economic, and socially positive spaces for occupants to enhance their comfort, productivity, performance, and social interactions.

The objectives of this research paper are the following:

- Improve productivity and sustain the mental well- being of occupants.
- Create an energy efficient interior environment with various strategies for cost saving and eliminating excessive energy consumption.
- Reduce the environmental and health impacts with the proper selection for furnishing and finishing materials.
- Aim for longevity and flexibility in the design.

2 The Case Study

The chosen site is the main lobby of the institutional building: The British University in Dubai (BUiD). The university is the first research-based, postgraduate university in the middle east with programs offered in various fields such as Business, engineering, and informatics. Additionally, the university is working in partnership with leading

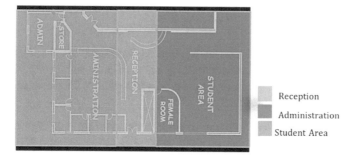

Fig. 1 Zoning Plan

UK universities. Building facilities include classrooms, auditorium, student area, the writing center, and Labs.

Site Analysis
The chosen site is located in Dubai International Academic City, with coordinates of 25.1266°N, 55.4109°W. BUiD was established back in 2003 as one of many universities all in the Academic City, which was named as a serving purpose indicating the presence of all the educational institution in the area. The British University is located in an open area with great exposure to natural sunlight and wind all year. Although the city is considered a relatively new one, the educational institutions have transformed the area to an interactive one. Figure 1 demonstrates the orientation of the building and the sun path in the location as well as the site boundaries and main surroundings. The paper will be investigating one section inside the university, which is the main lobby on the first floor that includes several spaces.

Existing conditions
Details of the existing site conditions will be listed such as dimensions of the spaces, the finishing of the walls, ceiling and floors, lighting fixtures, and materials of the loose furniture.

The total built up area of the chosen space is about 460 m², which include the following zones:

- The reception zone which includes the reception counter and two waiting areas with approximately total area (96 m²).
- Student's zone includes activities area, female room, and a store with approximately total area (120 m²),
- Administration zone includes open plan office area in addition to six closed offices with approximately total area (195 m²).

2.1   Drawback on Existing Conditions

Administration and reception area has no direct daylight and natural ventilation hence these areas only depend on artificial lighting and ventilation and it's not possible for the occupants to use the space or function properly without the presence of one or both of the previously stated.

- The design of the student area does not achieve the calm relaxing atmosphere, the used color tones for the furniture and walls are strong and alarming, the lighting is bright white and excessive which is not required for this type of space. Additionally, the furniture has non-eco-friendly materials with synthetic fabrics and tables with long processing and energy consuming manufacturing.
- The reception design lacks the welcoming and healthy atmosphere vibe in addition to the excessive white lighting where it's not needed. The waiting area furniture is a non-eco-friendly option since it's made of long processed plastic which requires high energy consumption and it's harmful for the environment and occupant's health.
- The lack of sufficient nature elements in all the spaces which is significant to bringing the occupants closer to nature and providing a better indoor air quality therefore enhancing their mood, productivity, and performance.

3   Methodology

The paper will use several analysis methods through computer programs for the existing conditions of the chosen space and the proposed design. Three factors will be considered during the design concept stage, namely environmental, economic, and social impacts where these impacts are directly related to achieving the desired solutions for a sustainable interior design. These factors are referenced by the ASHRAE standard which is an international standard advancing the well-being of occupants through sustainable technologies in the built environment. The interior finishing materials will be evaluated through LCA (life cycle analysis) to assess the material's quality in terms of environmental and health impact throughout its entire life cycle starting from raw material extraction until the final stage of disposal. Additionally, an energy evaluation will be conducted for the energy consumption in the space in order to aim at reducing the amount of consumption in the proposed design using more suitable strategies for lighting and ventilation. Furthermore, a comparison is conducted between the existing design and the proposed design in order to demonstrate the areas of improvement according to the used strategies of sustainability by DesignBuilder Software.

4 Proposed Design

4.1 Sustainable Design Development

Sustainability is a concept that holds within it broad views and consideration of many other concepts that affect the human civilization and defines its interactions and relationships with ecosystems and surrounding environment, of which led many scientists to agree on avoiding entitling it to one specific meaning (Ramsey 2015). Nonetheless, it can be concluded that sustainability is the result of the environmental concern and sense of responsibility towards the planet as whole. The concept of sustainability can be simplified into "meeting the needs of the present without compromising the ability of future generations to meet their own needs" (Portney 2015, p.4). Sustainability as whole discusses sustainable ecosystems, sustainable energy, society and economy, sustainable development, environmental protection, conservation / preservation of natural resources (Farley and Smith 2020). The main distinguish is that sustainability tends to be a long-term dynamic process of all the other concepts regarding the three Es: Environment, Economic and Equity, where these three pillars of sustainability must be met and fulfilled simultaneously through protection, promotion, achieving growth and development for all at once, where no pillar is prioritized over the other, and no pillar is overshaded and neglected or sacrificed for the flourish of another. The maximum potential and yield of sustainability have yet to be achieved as the international community still lacks the awareness and perception of the importance of sustainability and its different application in all lives. The following sections discusses individually the social, environmental, and economic aspects of adopting sustainability in interior designs.

Economic Aspect

Another essential pillar of sustainability is the economy and prosperity of humans' developments. The economic sustainability balance between the economic growth and savings and the environmental considerations and restrictions of natural resources; in order to achieve a sustainable economic growth without affecting the environmental aspects of it. The active approach is achieved through the proposed lighting system, that is 40% more efficient than the baseline (current) and is still environmentally friendly with LEDs technology. In addition, the proposed fan power in the HVAC system is more efficient and less energy-consuming, and has a smart control system that changes the fan speed according to the cooling load at that instant, which result in savings in electricity bill. The passive approach is through applying sealant on the gaps that decreases the air leakage inside the building envelope, which decreases the cooling load on the HVAC system. Additionally, decreasing the air leakage from 10 $m^3/m^2/h$ which is the maximum leakage allowable from Dubai Green Building Regulation to 5 $m^3/m^2/h$ which is the recommended allowable air leakage for offices and buildings according to the UK standard - ATTMA, and in addition to an extra layer to be added to the interior walls for thermal insulation promotion, decreasing the thermal losses to outdoor (unconditioned area).

Social Aspect

The social benefits of implementing sustainability in interior design are related to improvements of occupant's quality of life, mental and physical health. These benefits can be perceived at three main topics, namely health, comfort, and satisfaction. These outcomes can be interrelated, yet they have different employment methodologies. The proposed design solutions investigated the various outcomes of health, comfort, and wellbeing which are associated with the presence of sustainable design approaches. The built environment can have negative and positive impacts on the quality of life of occupants. Some approaches such as increasing occupants' control over indoor environmental conditions, access to natural views and daylight, higher connection with nature is likely to create positive impacts on occupant's health and wellbeing which are closely related to their levels of performance and productivity (Benjamin 2018). Environmental Aspect: The environmental aspect is an important pillar of Sustainability, which is achieved through protection and preservation of different ecosystems, with controlled and safe interactions with human development processes, while still achieving a reasonable conservation of natural resources and utilizing them with respect to the present generation needs and avoid any excessive exploitation that might lead into accelerating their depletion, resulting in shortage of natural resources for the future generations. Generally, environmental sustainability was referred to as the "environmentally-responsible development" and the "environmentally-sustainable development" by the World Bank in 1992 when sustainability was introduced officially and implemented, where until 1995 the term was officially developed as environmental sustainability (Moldan, Janoušková & Hák 2012). The intention of adopting a sustainable interior design is to improve humans' welfare, lifestyle and living conditions by protection the environment and its resources of raw materials and minimize the human waste (i.e., residential, agricultural, construction/commercial, …, etc.) to align it with the available natural sinks (soil, water, atmosphere) without exceeding its capacity that might lead to a reverse impact, resulting in irreversible harmful impacts on humans and different ecosystems in terms of health and safety, instead of serving its main purpose of elimination of the waste. The natural sinks are in this case consisting of two subsystems: the emitted pollution and waste absorption (Moldan, Janoušková & Hák 2012). On the other hand of the cycle, there is the sourcing side consisting of the renewable and non-renewable sources. These two subsystems are studied and analyzed of which renewable resources are promoted, investigated, and invested in with time and money, and the harmful impacts of the different stages of using nonrenewable resources are minimized, mitigated, and compensated for. Consequently, all of these approaches will lead to minimize the exploitation natural materials (Aluminum, wood, metal, …, etc.) will bring back the biological diversity and natural geometry, resulting in a clean fresh air and clean, flourished water and land resources, enhancing the human health and development. In addition, it will also improve the overall integrity of the biosystems, sustaining all the options for future generation, and maximizing the options available as whole (Ones & Dilchert 2012).

4.2 Plan Layout

Figure 1 bellow shows the space zoning for the new layout plan. The spaces were allocated according to the requirements of each space function. The administration area was placed in the area where more daylight can penetrate into the space reducing artificial light consumption as the office area requires an adequate amount of light in order to allow the occupants to properly perform their tasks. The student area was allocated in a place with less day lighting as it would be sufficient for this type of function which is mainly relaxing, socializing, and playing games. The reception area is at the center of the lobby facing the main door in order to facilitate the guidance for new visitors.

4.3 Design Specification

Walls, Floors, and ceiling finishes

The following are the finishing materials used for the walls, ceiling and flooring using sustainable approach specifications.

- Concrete texture wall painting with low VOC
- Modular Glass partition pre-assembled units with easy installment and removal to allow for an easy reallocation.
- Green wall for a welcoming, warm environment in the reception area in addition to improving the indoor air quality due to the natural plantation in the wall.
- Sustainable wood flooring boards made from recycled plastic and wood powder, non-hazardous chemicals used for installing in addition to easy installation and cost-efficient qualities.
- Marmoleum floor covering, which is also knows as linoleum, it's made from natural raw materials such as wood & cork dust and linseed oil. It's a highly durable and eco-friendly product with manufacturing that uses 100% green energy.
- Gypsum false ceiling which is recyclable and does not contain or requires the use of toxic substance for manufacturing and installment (Tables 1 and 2).

Table 1 Wall specification

Item	Recyclability	Sustainability advantages
Glass partition	100% recycled aluminum	Recycable, Durable, Reusable & flexible Sound absorbing and allow natural sunlight
Green wall	100% recyclable	Durable and can be broken down to be reused/ Stabilize moisture level in air (higher air quality)

Table 2 Material specifications

Item	Material specifications	Sustainability ADV
Seats	Cotton blend fabric	OCS certified (Organic content standard) organic cotton. Renewable, durable, recyclable & bio-degradable fiber
	Plywood	Biodegradable. Long life & can be recycled to reduce deforestation
	RECYCLED plastic	100% recycled
Stools	RECYCLED plastic shell, sled base	41.23% recycle content
Counter	WOOD veneer laminate finish	Durable (stain resistant) Biodegradable
Desk	WOOD veneer laminate finish	Durable (stain resistant) regularly renewable recyclable, biodegradable

Fig. 2 Section AA

Fig. 3 Section BB.

Furniture Selection and Specifications

Sections
Figures 2, and 3 demonstrate the proposed design simulation for the spaces which includes 3d perspectives and sections to demonstrate the height and interior elevations of the spaces.

5   Analysis and Discussion

In this paper, the economic and environmental aspects have been studied and analysed separately using simulation software Design Builder and LCA analyser, respectively. Two energy models have been created: the baseline and proposed; in order to carry

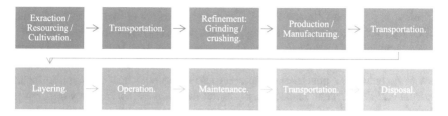

Fig. 4 Materials' main life cycle stages

Table. 3. Output data of pyrolysis and gasification of plastic waste

Process emission	Pyrolysis	Gasification
CO_2	150 kg	181 kg
CO	0.07 kg	0.45 kg
NO_x	0.6 kg	0.49 kg
SO_2	3.36×10^3 kg	0.091
PM (Dust)	9.52×10^3 kg	0.18
Solid Waste	65 kg (sent to WTE)	130 kg (sent to WTE)

out a simple comparison between the initial state of the case study and the proposed interior design in terms of environmental impacts of chosen materials (i.e., finishes, flooring, furniture, and walls), and in the economic savings from the lighting and HVAC analysis.

5.1 Life-Cycle Analysis

LCA analysis was carried out in this paper to determine the environmental impacts of the proposed materials over their lifecycles, in terms of resourcing, manufacturing, transportations and operation (Krishna et al. 2017). In addition, it analyses the end-of-life cycle for the material and the disposal method, whether it consists of recycling, reusing or burial and demolishing. The flow chart below (Fig. 4) shows a simplified the materials' main life cycle stage.

The environmental performance regarding the life-cycle assessment of the proposed material were studied and analyzed as shown below (Table 3).

5.2 Lighting Solution

Light-Emitting Diodes (LEDs) are an efficient lighting system that is made from semiconductors that emit light in the case of an existing electrical current flowing through the integrated microchips, causing the emission of visible light used for different application (Zhuang 2018). These lights mainly utilize the technology of LEDs for energy efficiency and savings as it uses at least 75% less energy compared to the consumption of Incandescent and Fluorescent lights. It reduces life cycle impacts and end-of-life consideration and emission and thermal release where it emits very low to no heat in its operation compared to 90% and 80% heat emission in Incandescent and Fluorescent, respectively (Gómez and Izzo 2018). Not only that, but its durability and longer lifetime of 3 to 5 times than other light. Moreover, the quality of LEDs lights is also evident in its intensity and flexibility in direction of emitted light without any necessary diffusers or reflecting medium without losing the light half through the way from the fixture (Tuenge 2013). It also comes with a wide range of colour options, which have different psychological impacts and applications. Three LEDs lights were used: Integrated LED pendant light, Integrated Grille Spotlight Downlight LED Embedded, and LED Strips Profile Aluminum Channel Aluminum Linear Lighting Profile. All three proposed lights are 100% recyclable and are manufactured from and by eco-friendly materials and processing. The proposed lighting layout is targeting 40% saving on the baseline in terms of energy consumed that is assumed to be following ASHRAE 90.1–2007, which is the minimum lighting schedule followed in UAE.

5.3   Energy Analysis

For energy analysis, two simple energy models have been created in Design Builder Simulation Software: the baseline and proposed; in order to carry out a comparison between the initial state of the case study and the proposed interior design in terms economic savings from the lighting and HVAC analysis. The four main changes that are made between the baseline and proposed cases are the following:

1. Lighting Schedule.
2. U-Value of interior wall and Insulation Thickness.
3. Allowable air leakage through the envelope in $m^3/m^2/h$.
4. FCUs fans rated power in kW.

The changes made are all summarized in the Table 4. The table shows 7.4% savings in the cooling load, which is the result of decreasing the air leakage from 10 $m^3/m^2/h$ as per DGBR minimum compliance to 5 $m^3/m^2/h$ as per the recommended from ATTMA. In addition, the U-value of the walls was decreased from 0.704 to 0.403 W/m^2. K, Additionally, with increasing the efficiency of the fans and decreasing the power needed, less thermal heat losses were released from the fan is to the air stream. Comparisons of annual energy consumption between the base case and proposed solution are shown in Fig. 5.

Table 4 Comparison between base case and proposed solution in terms of saving

End use	Baseline	Proposed	Savings	Ratio	
	[kWh/a]	[kWh/a]	[kWh/a]	[%]	[%]
Cooling	31,064	26,599	4,465	7.4%	86%
Fans-interior	5,564	2,348	3,216	5.3%	42%
Interior lighting	12,913	7,748	5,165	8.5%	60%
Receptacle Eqp	10,889	10,889	0	0.0%	100%
Total	60,430	47,584	12,846	21.26%	78.7%
EUI [kWh/m^2/year]	14.6	11.5	3.1		

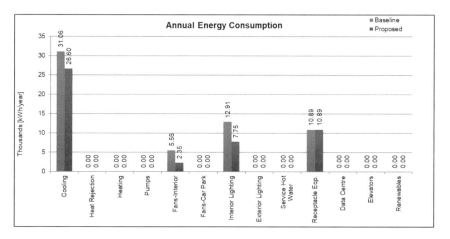

Fig. 5 Annual energy consumption: proposed case Vs Base case

6 Conclusion

The paper aimed to achieve various solutions through redesigning the spaces of the main lobby of the BUiD University using sustainable interior design approaches which are related to improving the occupant's quality of life, reducing health and economic impacts that's imposed by the traditional design while also considering the environment which is a closely related to human existence and survival. Various strategies were used in order to achieve the sustainability goals in interior spaces. These strategies included the proper analysis and selection for materials, products, lighting and colours in addition to the nature elements and day lighting that helps occupants be more connected with nature and outdoors.

References

Benjamin, H. (2018). Human health and well-being through LEED. https://www.usgbc.org/articles/human-health-and-wellbeing-through-leed. Accessed 19 Mar 2022

Farley, H. M., & Smith, Z. A. (2020). *Sustainability: if it's everything, is it nothing?*. Routledge.

Gómez, C., & Izzo, L. G. (2018). Increasing efficiency of crop production with LEDs. *AIMS Agriculture Food, 3*(2), 135–153.

Hayles, C. S. (2015). Environmentally sustainable interior design: a snapshot of current supply of and demand for green, sustainable, or fair-trade products for interior design practice. *International Journal Sustainability Built Environment, 4*(1), 100–108.

Krishna, I. M., Manickam, V., Shah, A. and Davergave, N., 2017. Environmental management: science and engineering for industry. Butterworth-Heinemann.

Linhares, T. B. & Pereira, A.F. (2017). Sustainable Buildings and Interior Design. SBDS+ ISSD, pp.82–87.

Moldan, B., Janoušková, S., & Hák, T. (2012). How to understand and measure environmental sustainability: indicators and targets. *Ecological Indicators, 17*, 4–13.

Ones, D. S., & Dilchert, S. (2012). Environmental sustainability at work: a call to action. *Industrial Organizational Psychology, 5*(4), 444–466.

Portney, K. E. (2015). *Sustainability*. Cambridge: MIT Press.

Ramsey, J. L. (2015). On not defining sustainability. *Journal of Agricultural and Environmental Ethics, 28*, 1075–1087.

Reuter, M. A., & van Schaik, A. (2015). Product-centric simulation-based design for recycling: case of LED lamp recycling. *Journal Sustainable Metallurgy, 1*(1), 4–28.

Tuenge, J.R., Hollomon, B., Dillon, H.E. & Snowden-Swan, L.J. (2013). Life-cycle assessment of energy and environmental impacts of LED lighting products, part 3: LED environmental testing (No. PNNL-22346). Pacific Northwest National Lab. (PNNL), Richland, WA, United States

Zhuang, Y., Hua, L., Qi, L., Yang, J., Cao, P., Cao, Y., Wu, Y., Thompson, J., & Haas, H. (2018). A survey of positioning systems using visible LED lights. *IEEE Communications Surveys Tutorials, 20*(3), 1963–1988.

Summarising a Twitter Feed Using Weighted Frequency

Zina Ahmed Abohaia and Yousef Mamdouh Hassan

Abstract Data is growing exponentially every day, with 500 million tweets sent on Twitter alone (Desjardins 2021). Twitter feeds are long, take time to understand, are multilingual, and have multimedia. This makes it difficult to analyse in its raw form so the data needs to be extracted, cleaned, and structured, to be able to be used in research. This paper proposes summarising twitter feeds as a manner of structuring them. The objectives we sought to achieve are: (1) Use the Twitter API to retrieve tweets successfully, (2) Efficiently detect the language of text, and tokenize it to then analyse their content (in its language), (3) Use live tweets as the input instead of a database of tweets, (4) Create the interface as a plugin to make it accessible for computer scientists, and others, alike. We also aimed to test whether using weighted frequency to construct summaries of tweets would be successful, and by conducting a survey to test our results, we have found that our program is seen to be useful, accessible, and efficient at giving summarizations of twitter accounts. Weighted frequency also proved to be good at summarising text of any language, inputted.

Keywords Natural language processing · NLP · Twitter · Summarization · Weighted Frequency

1 Introduction

Data analysis is a required tool since 90% of the entire world's data was created only in the past 2 years (Marr 2018) where, according to Forbes (2018), "2.5 quintillion bytes of data are created every single day". Social media is also getting bigger and hosting more data, and on Facebook alone, there are 2.6 billion people who are monthly active users (Tankovska 2021). As technology and social media advance more, more data will be created, and therefore the need for data analysts and data analysis tools to process this data efficiently will grow.

Z. A. Abohaia (✉) · Y. M. Hassan
The British University in Dubai, Dubai, United Arab Emirates
e-mail: 20194621@ug.buid.ac.ae

K. Al Marri et al. (eds.), *BUiD Doctoral Research Conference 2022*,
Lecture Notes in Civil Engineering 320,
https://doi.org/10.1007/978-3-031-27462-6_17

That's because most of the data created on the internet is in its raw form, i.e., needs to be analysed to be useful in research. In reality, only 0.5% of all data available online is being analysed (Wassén 2018), therefore more tools that clean, structure, and summarise are needed. Since 500 million tweets are sent on Twitter alone (Desjardins 2021), the question we ask is, can Twitter accounts be successfully summarised?

Twitter accounts have many issues that plague the Natural Language Processing (NLP) field such as: having multimedia data, data in multiple languages, slang and sarcasm. Other issues that would face non-computer scientists would be using the Twitter API itself. There are many methods that have been used to structure, and use data retrieved from Twitter but Summarization, particularly using weighted frequency, has not been tested. Summarising a twitter account would give researchers the chance to get a gist of what an account is about, as well as use it as a form of structuring this great amount of data, not only cleaning it but also getting only what is important from it. Not only that, it is also a way to store the data effectively.

The Objectives of our project are (1) Use the Twitter API to retrieve tweets, (2) Efficiently detect the language of text, and tokenize it to then analyse their content, (3) Use weighted frequency to build sentences, and output the most relevant points, (4) Use live tweets instead of a database as input to the algorithm, (5) Create the interface as a plugin so it is easily used by users to increase accessibility.

The research questions we explore are whether a plugin interface would be accessible for casual users, professionals, and researchers alike, as well as if we can effectively summarise a twitter feed using weighted frequency.

2 Literature Review and Comparative Analysis

Each of these different papers use the available information for a specific purpose, and all have in common techniques of data pre-processing, and analysis. We have identified a gap in the usage of Twitter simply to make a tool for the people, to be used by the people. Whether those people are computer science researchers, non-computer science researchers, or normal users, we want to provide all a simple tool to structure unstructured data, and get a general idea of Twitter accounts easily and quickly.

Casteleyn et al. (2009) experimented using Facebook for market research. Using a social theory, Dramatism, for evaluation of the content, instead of a purely mathematical evaluation, they aimed to structure the unstructured data on Facebook to: 1- understand customers 2- the nature of the market. They wrote a purpose statement then gathered terms related to it from the data available. Using those results, they rated the text using Dramatism, and concluded that it is nothing but a "heuristic model", and that their study might help researchers in understanding bias in certain countries of the world. However, academic researchers no longer have permission to use data from Facebook, and making a database instead of using live data does not give as accurate results.

Meyer et al. (2021) improved web mining, and data structuring methods for the purpose of analysing unstructured data, extracted from Twitter and Wikipedia, to use them in food crisis management. They extracted both German and English text and used an application to get and save content from Twitter which they filtered using rules and keywords. However, cleaning the text, and finding relevant tweets, still proved to be a great challenge. Taking count of the number of page views of Wikipedia pages to use it in analysis was novel but did not prove helpful.

Stieglitz et al. (2018) reviewed existing literature to identify challenges researchers face in Data Extraction and Analysis, as well as the methods they use. They developed a four stage framework consisting of Discovery, Tracking, Preparation, and Analysis to deal with social media analytics. They found that the most challenges are found in managing the high volume of data and suggested interdisciplinary solutions to analyse data qualitatively, as well as quantitatively simultaneously. Other problems and solutions included the storage of data, and its visualisation but there was no mention of its summarization.

Adedoyin-Olowe, Medhat Gaber and Stahl (2013) reviewed available data techniques such as opinion mining, data gathering, and summarization. Particularly, opinion summarization, which uses polarity to achieve its purpose. This requires the rating of every opinion which is tedious work, and not effective. There was no mention of weighted frequency, and the techniques were only listed with their potential uses, with no comment on their relevant effectiveness.

Bessagnet (2019) developed a generic framework, to perform a comprehensive analysis of French and English tweets, which consists of: preparation and validation, first analysis, multidimensional Analysis, and summarization. The summarization stage was not clear but it seemed to be a method of information retrieval, where rules are used to validate, parse, and tokenize data.

Cheong and Cheong (2011) utilised Twitter to see whether live tweets may help during floods. They used the Twitter API to scrap the tweets, graph theory to analyse it. The data was then displayed as a network where they confirmed the presence of authorities providing help on social media, and found that the information provided online was more general rather than critical.

As can be seen from the previous literature review, summarization using weighted frequency, has not been explored as a data mining technique, and that the researchers were limited to analysing data only from the languages they spoke, and not all languages (Table 1).

Table 1 Summary of Papers

Study	Purpose	Sources used
(Casteleyn et al. 2009)	Use the theory of Kenneth Burke, a critical technique called Dramatism, which emphasises the motives of humans, to analyse the nature of social media	Facebook groups
(Meyer et al. 2021)	Check if Web mining and social media analysis can be used for food crisis communication, which is crucial to successful crisis management	Twitter, and Wikipedia
(Stieglitz et al. 2018)	Highlight challenges, potential solutions, as well as extending an existing framework on social media analytics	Existing literature that uses mainly Twitter
(Adedoyin-Olowe et al. 2013)	Survey different data mining techniques of social networks	Existing literature that uses Twitter, Facebook, LinkedIn, Google+
(Bessagnet 2019)	Present the use of social data analysis by non computer science researchers, using different tools	Twitter
(Cheong and Cheong 2011)	To determine active players, and their contribution to critical information during the 2010–2011 Australian floods	Twitter

3 Project Methodology

3.1 Use Case Diagram

There are two main use cases, as we can see in the diagram below. The first one is when the user downloads the extension, and this is required for the user to gain access to the extension and its features. The second use case is when the user is on a webpage, and clicks on our extension, leading to an analysis of the URL and including the displaying of results (Fig. 1).

Fig. 1 The Good Analyst
use case

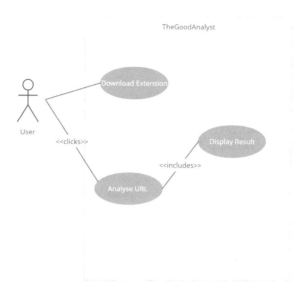

3.2   Implementation

We used the Twitter API, in python, to extract tweets live from Twitter. The program sends Twitter a request and the API returns a response, retrieving in our case data and public tweets. To detect the language the tweets are in, and get the stopwords unique to every language, we used the python library langcodes. After the tweets are retrieved, we tokenized them into sentences, then into words, using the python libraries regular expressions and NLTK. Then we calculated the weighted frequency of occurrences, replacing each word with its weight.

Finally, for every tweet, the weights of all its words were added up, and the tweets with the highest values were outputted as results. We used HTML to create the base to the user interface and CSS and JavaScript to make it eye friendly and easy to use, and used a python library called "eel" to create python apps with HTML via hosting a local webserver and calling functions from python via JavaScript.

4   Evaluation and Results

4.1   Summary of Survey Results

Using a 11-question survey, we evaluated our project with the help of 58 participants. The questions were written based on the questions, and recommendation of Davis (1989). We aimed to know whether users will find the extension useful, accessible, easy to understand, eye-friendly, and we also had them review three examples of inputs and outputs to our program. Some of our findings are:

- More than three quarters of the participants agreed that they find it easy to use an extension.
- Almost 9 out of every 10 of the participants found the features of our extensions easy to understand from the summary included at the start of the survey.
- 9 out of every 10 participants say that our extension achieves its purpose, and that it would help them understand what a Twitter account is about.
- Accessibility also proved to not be an issue, with 85% of the participants agreeing that they find extensions accessible.
- Three quarters of the participants would be interested to use our tool.
- Half the participants said that our Results page was not eye-friendly.
- Combining the results for our three examples and averaging them, we find that around three quarters of participants agree that our results are a good summary of the accounts used in the examples.

We were more than delighted with the number of people who found the project accessible, easy to understand, and useful, proving that, albeit on a small scale, this tool would be beneficial to people.

We predicted more people would find the results agreeable, but since the results are outputted and changed with every new tweet, I can see how an output ten days ago may be more agreeable than the output now. Either way, the result for that was still within our accepted range.

We did not expect half the participants to find the results page eye-friendly, and the other half to find it not, since we began by trying to use perfectly complementary colours. We received some feedback, and therefore, we have changed our results page.

4.2 Survey Motivation

We have carried out this survey for the following purpose: to evaluate potential users' opinions on extensions, our algorithm, and the potential usefulness of our tool. This information helps in:

- Testing and improving our algorithm
- Testing the likeability of our results page, and improving it
- Having an approximation of how many people would use our tool if we did deploy it to market
- Making sure it serves the purpose it was made for

4.3   Results and Discussion

As seen in Fig. 2 and 3 above, these are two examples of our code's output. We have
kept the links to make sure that no critical information was lost, and that after storage
and reviewal of it, the content makes sense and is in context. The results successfully
weighted tweets in different languages, and outputted the tweets with the highest
weights, regardless of language.

As seen in the summary of the survey results, three quarters of participants found
the result of the code a good summary of the accounts used in the examples, and 9
out of every 10 participants said that it would indeed help them understand what a
twitter account was mainly about.

For our last question, we inputted to our program a News Twitter Account. It should output
the most important news/tweets in the last 100 tweets. This was the output for the 6th of
June:
https://t.co/zKtsGT2fne#UAE airlines ramp up UK services, but with strict #COVID19
guidelines https://t.co/M42205FwTQWATCH: Where is the UAE visa number on my
passport?

Find out how to get in#Food #Foodiehttps://t.co/sh5cIpS383#Bollywood actor
#SunilGrover on filming 'Sunflower' during #COVID19.

Then read this https://t.co/e85RunBO0Y#SaudiArabia: Man dangles from #Riyadh
pedestrian bridge.

The bus overturned after colliding with the van coming from the opposite direction
https://t.co/yFoFaSToZz#COVID19 and #India's 7-year itch with the #BJP.

Confidence in the outlook for oil demand continues to grow as accelerating vaccinations
https://t.co/h7wWRZaOGDDespite measures by authorities, surge seen in domestic
violence in #Egypt amid #COVID19 pressures.

@sheela2010 Indian academic, author to #GulfNews https://t.co/qFd1rkfyg]Best
breakfast in #Dubai.

Al Hosn Green Pass: What is it and how can it be used?https://t.co/84odb9w03PUAE
expats: Time to remit?

Fig. 2   The last question of the survey: the results of using the program on a news account

ربما تكون أحد أكثر أنواع الزبدة انتشاراً في الوقت الحاضر هي "زبدة الفول السوداني"، و التي نتمتّع بها

جميعاً, كباراً و صغاراً. للمزيد: https://t.co/MWKRSbZazN https://t.co/mHaCQRI.zuK

تطبق الشروط و الأحكام, عروض المتجر الإلكتروني الذكي الحصرية وصلت! للمزيد:

https://t.co/QPflSoD3Yo https://t.co/iqEdJrwptD

قم بتنزيل المتجر الإلكتروني الذكي و اطلب الآن! تطبق الشروط و الأحكام. للمزيد:

https://t.co/pM3QkzzzkQ

يسري العرض من 28 مايو و لغاية 6 يونيو في كافة فروع تعاونية الاتحاد تطبق الشروط و الأحكام.

للمزيد: https://t.co/pM3QkzRaJg

ابدأ التسوق الآن! تطبق الشروط و الأحكام. للمزيد: https://t.co/pM3QkzRajg

For more: https://t.co/pM3QkzRaJgT&C apply.

https://t.co/e3XFYAV6lXClick and collect your favorite products now from the following branches of Union Coop: Al Towar, Al Aweer, Mirdif, Al Barsha, Umm Suqeim.

https://t.co/YV9QKcPBAC https://t.co/O24QUJbZAZPerhaps, one of the most popular spreads we have today is 'peanut butter', which by all means is enjoyed equally by old and young.

Fig. 3 A multilingual example of results, Union Coop's Twitter account

5   Conclusion and Future Work

Data is growing at an exponential rate, and the extraction, and analysis of it are proving to be a problem. In this paper we have seen that the analysis of twitter Feeds of any language, as well as multilingual text, is possible using summarization by weighted frequency. Our survey showed that our program was found to be accessible, easy to use, and useful. The standardisation of this method could be useful to researchers. For our future work, we aim to make an academic twitter account to access a greater number of tweets, and thus provide more accurate results. Furthermore, the code is available at (Summarizing A Twitter Feed Using Weighted Frequency 2022).

References

Abohaia, Z., & Mamdouh, Y. (2022). Summarizing A Twitter Feed Using Weighted Frequency. Github. https://github.com/ZA8422/Summarizing-a-Twitter-Feed-using-Weighted-Frequency-.git. Accessed 21 Aug 2022

Adedoyin-Olowe, M., Medhat Gaber, M., & Stahl, F. (2021). A survey of data mining techniques for social network analysis. *Journal of Data Mining & Digital Humanities.* https://arxiv.org/abs/1312.4617 Accessed 6 June 2021

Bessagnet, M. (2019). A generic framework to perform comprehensive analysis of tweets. In: *7th International Workshop on Bibliometric-enhanced Information Retrieval.* https://hal.archives-ouvertes.fr/hal-02414037. Accessed 6 June 2021

Casteleyn, J., Mottart, A., & Rutten, K. (2009). Forum - how to use Facebook in your market research. *International Journal of Market Research, 51*(4), 439–447.

Cheong, F., & Cheong, C. (2011). Social media data mining: a social network analysis of tweets during the 2010–2011 Australian floods. In: PACIFIC ASIA CONFERENCE ON INFORMATION SYSTEMS (PACIS) 2011 proceedings. https://aisel.aisnet.org/pacis2011/. Accessed 6 June 2021.

Meyer, H., & C., Hamer, M., Terlau, W., Raithel, J. and Pongratz, P. (2021). Web data mining and social media analysis for better communication in food safety crises. *Int. J. Food System Dynamics, 6*(3), 129–138.

Most used social media 2021 | Statista. Statista. https://www.statista.com/statistics/272014/global-social-networks-ranked-by-number-of-users/. Accessed 9 Feb 2021

Marr, B. (2018). How Much Data Do We Create Every Day? The Mind-Blowing Stats Everyone Should Read. Forbes. https://www.forbes.com/sites/bernardmarr/2018/05/21/how-much-data-do-we-create-every-day-the-mind-blowing-stats-everyone-should-read/?sh=5bdf349760ba. Accessed 21 May 2018

Stieglitz, S., Mirbabaie, M., Ross, B., & Neuberger, C. (2018). Social media analytics – challenges in topic discovery, data collection, and data preparation. *International Journal of Information Management, 39*, 156–168.

Wassén, O. (2018). Big Data facts - How much data is out there? | NodeGraph. NodeGraph. https://www.nodegraph.se/big-data-facts/. Accessed 1 Jan 2020

Preventive Maintenance Using Recycled Asphalt: Review

Aishah H. O. Al Shehhi, Gul Ahmed Jokhio, and Abid Abu-Tair

Abstract The article addresses the importance of using recycled asphalt with the integration of road maintenance procedures in the road network. The road network is considered the main element of any national infrastructure development plan. The research aims to study and highlight the using recycled asphalt as a suggested sustainable method for road maintenance procedures. Therefore, the study elaborates on the historical use of recycled asphalt, its advantages, and disadvantages. Besides that, maintenance process categories to ensure the suitable type that helps provide the best quality of the network. Since roadway pavement assessment is based on quality as well as different characteristics parameters such as rutting, cracking, pavement quality Index, and roughness in addition to other parameters. The primary method used in this study is the review of existing literature. It can be concluded that the use of recycled asphalt in road maintenance can become a viable and sustainable alternative to current road maintenance practices.

Keywords Recycled asphalt · Roadway maintenance · Pavement Parameter · Condition Index

1 Introduction

Roadways are one of the significant important elements in infrastructure because they are characterized simply as the point of interaction between societies and peoples. the country needs to create roadways so that citizens and visitors may travel easily and smoothly. Additionally, roadways and highways have been the main source by which whole economies and societies have emerged and developed over the years. They

A. H. O. Al Shehhi (✉)
The British University in Dubai, Dubai, United Arab Emirates
e-mail: ayshalshihi@gmail.com

G. A. Jokhio · A. Abu-Tair
Structural Engineering, The British University in Dubai, Dubai, United Arab Emirates

© The Author(s) 2023   189
K. Al Marri et al. (eds.), *BUiD Doctoral Research Conference 2022*,
Lecture Notes in Civil Engineering 320,
https://doi.org/10.1007/978-3-031-27462-6_18

also made a positive contribution to the distribution of ideas, cultures, languages, discoveries, goods, and services (Koch and Ksaibati 2010).

The history of roadways started with the Ancient Egyptians carrying log rollers used to build the pyramids with huge stones. In avoiding the friction between rocks and dirt, these rollers were strong. Later, the Egyptians invented the wheel, which was considered a transport tool to reduce the need to reposition the log rollers and reduce the need for human power in moving materials. As a result, of minimizing the contact area, they discovered the need to build a hard strong surface to reach their destination. This was the first step in constructing the roadway (Byrne BEng 2005).

On the other hand, Babylon built the first bitumen pavement in the early of 600BC. The new material was used to support the strength of the stone slabs to prevent any further collapse. The roadway was over 1 km in length and besides that, history witnesses the first roadway that linked the big catchment area was King Darius I of Persia, which was built over 60 years. This roadway is a transaction of a straight path that connects two points. Another brilliant road scheme was Chandra-gupta's in India, which took around 30 years to execute with 2400 km in length. The main aim of this road was to enhance the traveling period. Besides that, the concept of roadway has been expanding in British with coverage of 78,000 km throughout Europe by establishing 5,000 km roads (Rosyidi 2015).

Roman became famous for their abilities and skills as roadways constructors over five centuries. They began the road network system among 80,000 km of main roadways and 320,000 km of inner roadways. The roadways covered about 62,000 miles, which were made of natural materials of lime, stone, and concert which were considered binders. Via Appia was the oldest Roman roadway constructed in 312 B.C (Asphalt Pavement History I Washington Asphalt Pavement Association 2010).

2 Recycled Pavement History

The beginning of recycled pavement was dated to 1915, but a dramatic increase in the price of asphalt binder during the 1970s Arab oil embargo sparked renewed interest in asphalt recycling. RAP was originally used by the industry for its economic benefits. Providing RAP is just one way the asphalt industry is attempting to become more sustainable. It is economically, environmentally, and socially beneficial. Using RAP reduces the cost of virgin material, emissions of carbon dioxide, and use of non-renewable resources.

The triple-bottom-line approaches are becoming increasingly ingrained in all aspects of life. The pavement community will continue to emphasize the use of recycled materials. A few difficulties that recently frustrated the progression of higher RAP amounts in blend plans were the absence of rules connected with blend plan and handling just like the accessibility of field execution information to show how these combinations act in the field. While these obstructions were the case a couple of years prior, late exploration and examination show that RAP blends can have identical execution contrasted with virgin combinations (Transportation Research Board 2014).

Additionally, (Koch et al. 2013) adapted that RAP is the term given to eliminated and reprocessed asphalt materials. These materials are acquired when asphalts are taken out for remaking, remerging, or to get sufficiently close to covered utilities. Therefore, RAP is being utilized in roadway network with the functionality of base or surface material. Road network is a main recycling infrastructure feature in the USA. Not many individuals understand that pavement recycling interstates are among the world's top recycled materials. Around 80% of asphalt is being reused in the roadways. That is contrasted and just 28% of reused post-purchaser merchandise in the city's strong waste stream. As per industry specialists, the recycled asphalt industry is the world's new treasure. Every year, 73 million tons of RAP are implemented in roadway constructions, saving citizens nearly $300 million yearly. The volume of RAP is 13 times more than papers, 27 times more than glass, and 267 more than plastics. Most of the original asphalts are reused within the location of the roadway. It can be utilized in various this way decreasing the interest in asphalt concrete in new or reused pavement containing RAP. At the point when utilized in black-top clearing applications (hot blend or cold blend), RAP can be handled either as hot in place or cold in place.

2.1 Hot in Place Recycled Pavement

This type of recycled pavement has considered as a promising approach and suitable solution for pavement restoration due to the ability of utilizing in place of the roadways . According to Asphalt Recycling and Reclaiming Association (ARRA) defines it as an on-site, in-place technique for restoring deteriorated pavements while reducing the need for new infrastrutucre materials whichfix surface distresses that aren't caused by structural flaws. A new wearing course is applied after recompacting the RAP materials in a multi-step technique, whereas the virgin materials are combined with the recovered reclaimed asphalt pavement (RAP) material in a single-pass operation. This type of recycled asphalt provides the advantages of conserving elevations and overhead clearances while also being very efficicnally in monitoring and controlling traffic volume.Recoat stripped aggregates, re-establish crown and drainage, change aggregate gradation and asphalt content, and increase surface frictional resistance may all be done using this technique. This type of recycled pavement is done to a depth of 20 mm to 50 mm (3/4 in to 2 in), with a typical depth of 25 mm (1 in).The amount of RAP is often extremely high. Whereas 15–20% RAP is prevalent in hot mix recycling, 80–100% RAP is common in hot in-place recycling. It is not essential to calculate any combination gradation of RAP and virgin aggregates if 100% RAP is employed. In hot in-place recycled mix, air voids can be as high as 4%. In Canada, higher design air voids (up to 6%) have been utilized effectively in hot in-place recycling (Kandhal and Mallick 2017).

2.2 Cold in Place Recycled Pavement

Various methods of recycled asphalt have been used to restore and maintain pavements in the United States since the 1930s. As a consequence of an ever-present focus on cost-cutting, as well as a growing emphasis on "green practices," contractors have started using these sustainable processes more regularly (Lombardo 2018). It is defined by (AASHTO 1998) as the process with bituminous and /or chemical additives of existing HMA pavement without heating to produce a restored pavement layer. It is an on-site treatment for moderate- to low-volume roadways that do not have significant underlying structural with a repair distresses 2 in. to 6 in. into the existing pavement.

Instead, for roads with raveling, weathering, bleeding, corrugations, pushing, sliding, rutting, cracking, and small craters, this approach is recommended. This process eliminates damaged layers, leaving a crack-free layer that may be used to install a new HMA overlay or surface course on top of it. If the pavement requires more substantial repairs and an overlay, CIR should be considered. CIR begins by milling 2 to 6 inches of old pavement, which is then processed, combined with a recycling agent (emulsified or foamed), repaved, and compacted (Lombardo 2018).

2.3 Full Depth Reclamation

Infrastruture authorities either local of government are entrusted with rehabitaion nation's infrastructure as it matures. Full-Depth Reclamation (FDR), a sustainable engineering solution for pavement restoration, might be the answer for these authorities to ensure the best quality of infrastructure to taxpayers while also being responsible stewards of public funds. Its a type of rehabilitation in which an existing asphalt pavement and its underlying layers are recycled into a new layers. The process begins with the use of a road reclaimer to pulverize an existing asphalt pavement as well as a part of the underlying pavement layers . The crushed layer is generally evenly blended with an extra stabilizing agent such as Portland cement to make an enhanced, homogenous composition. Finally, the stabilized material is crushed into place by rollers. The base is now strong and secure, ready for a new rigid or flexible surface course. This manual covers project selection, design, building, testing, quality control, and the FDR with cement technique, among other topics. is a pavement rehabilitation procedure in which an existing asphalt pavement and its underlying layers are recycled into a new foundation layer.

This is especially beneficial in urban communities and sattlements, in order to have smooth access to residential and commercial driveways is important during construction. Reclaiming in-place materials enhances staging while simultaneously shortening the construction schedule and minimizing traffic disruption. Full-depth reclamation using cement can be done in a way that enhances the geometry of the road. This process is molded with new mix materials to reach the desired cross-section after the older asphalt is combined with the other pavement layers. At this stage, minor profile and superelevation changes, as well as roadway widening, can be undertaken (Reeder et al. 2019).

2.4 Shoulder Surfacing and Widening

Over time, the role of shoulders adjacent as base bond of pavements. An additional temonologies of modern roadway shoulders are to accommodate an increasing encroachment of traffic, expedite water runoff from travel lane pavement, provide additional spaces for construction and maintenance activities such as establish dynamic surraoundings that ensure happenancess and welfare through walking paths, bicycle paths or slow-moving vehicles and equipment lanes, to reduce edge stresses and corner deflections. There must be paved shoulders on all Interstate routes.

Paved shoulders are justified by enhanced and smoother traffic operations, as well as the expectation of higher pavement performance, extended life cycle of roadways, increased highway safety, and cheaper maintenance costs. TRB Special Report 214, "Designing Safer Roads - Practices for Resurfacing, Restoration, and Rehabilitation," and FHWA/RD-87/094, "Safety Cost-Effectiveness of Incremental Changes in Cross-Section Design - Informational Guide," both claimed lower accident rates. Even if pedestrians aren't always expected in rural highway corridors, it's a good idea to leave enough room for them and future demands. A minimum paved shoulder width of 4 feet is recommended in the American Association of State Highway and Transportation Officials (AASHTO) Guide for the Development of Bicycle Facilities (2011) to accommodate pedestrian activity along rural roadways; however, this width may not be appropriate on high-speed roadways. The shoulder should be made of the same materials as the mainline pavement to make construction easier, improve pavement performance, and save maintenance costs (Shoulder Sealing I Road Safety Toolkit 2022).

3 Research Questions

Most of the previous articles and studies emphasized one pillar of sustainability and determined how the roadway can survive without maintenance. Therefore, the following Chart. 1 illustrates the summary of the research aim, questions, and objectives.

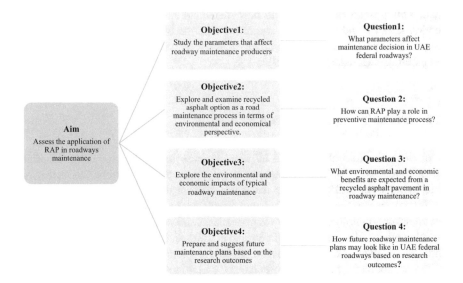

Chart. 1 Research main aim, objectives and questions summary, *sources*; author, 2022

4 Roadway Maintenance Process

Roadways can quickly deteriorate if they are not properly maintained regularly. Despite the initial investment in roads, this infrastructure deteriorates with time, necessitating not only regular roadway maintenance services to keep the existing roads in good repair, but also additional investments to improve and expand the system. Roadways will continue to deteriorate without this ongoing maintenance and development, requiring costly repairs or perhaps replacement after only a few years. A failing transportation infrastructure, soaring expenditures, and a large financial cost to the local economy and population arise from a lack of road maintenance services in a given location. Inadequate roadway maintenance impacts in reduction the level of services and sefety (Why Road Maintenance Is So Important 2019).

The purpose of maintenance is to keep the asset in good performance working order, not to enhance it. It's seen as a long-term investment that protects against costly maintenance and premature breakdowns. At the very least, maintenance extends the life of the transportation infrastructure by treating wear and degradation caused by traffic and the environment constantly. Minor fixes and modifications are occasionally included to address the source of issues and avoid unnecessary maintenance efforts. It also aims to maintain all highway infrastructure operational at the lowest possible cost and with the least amount of disturbance to the traveling public (Highway Maintenance Manual 2017).

5 Study Significance

The World Highways Organization set a target to have 70% of recycled infrastuture materials by 2020. This occurred in Europe where around million tons of recycled materials are used in the pavement that are producing valuable sustain scenarios at a lower rate of waste landfills. It should be noted that the number of materials that contribute to roadway pavements production such as gravel, and other aggregates of gravel is limited as raw materials and vital in different other industries as well such as building construction which means it shall be reserved and consumed with caution to ensure efficiency and sustainability (Balaguera et al. 2018). Therefore, this research serves a significant role in addressing the sustainability approach in different categories all at once by exploring new potential techniques for the roadway pavement industry. Studying the current situation of the pilot project in UAE and exploring the advantages and disadvantages of the RAP system according to UAE needs and potential will widen the possibilities of implementation of different recyclable materials in different other industries and encourage the development in pavement related businesses and industries to apply higher sustainable standards that will be reflected on all 3 pillars of sustainability. Previose researchs were done based on theoretical aspects which discuss disadvatanges of using recycled pavements however this research is the first on a national level that covers the main roadway which links high populated emirates with high traffic volume, especially on weekends. The results and outcomes will be taken into consideration for future policies and frameworks since the researcher already works at the Ministry of Energy and Infrastructure.

6 Methodology and Approach

The methodological Chart. 2 below presents data collection methods with an empirical perspective of qualitative and quantitative.

The research case study selection is Dibba – Masafi E89 with deep information such as length, traffic volume, type of maintenance, and others gave the author to analysis the need of maintenance through site experimental tests, LCCA, and Road Assets Management System. A comparison between the proposed maintenance

Chart. 2 Data collection, *source*: author, 2022

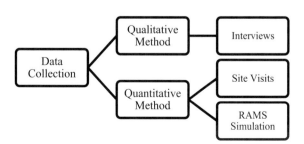

procedure of recycled pavement and current maintenance procedures in terms of environmental and economic impacts contribute with The findings and outcomes set the foundation for future maintenance.

For the qualitative method, the author holds a group of interviews with engineering experts from government authorities and the private sector. The objective of the interviews is to have clear comprehensive data and knowledge about the following:

- Type of maintenance process.
- Simulation tools that support the use of the recycled pavement.
- Group of parameters that affect the performance and quality of the roadway.
- Work producers of implementation recycled pavement with comparasion of duration.

Furthermore, site visits to the research case study to understand the following:

- Identify the roadway as location, and pavement layers.
- Kind of lab tests used to ensure the strength of the new asphalt.

While the second method of quantitative is a simulation which is described through RAMS which is considered as a dashboard for displaying road status information based on auscultation parameters, employing graphs, histograms, to illustrate the data. It produces group of reports probably difficult to list since they can be formed by a single or combination of parameters, however, the database can be graphically, numerically presented as well as mapped to indicate different topics such as bridges, rock slopes, assets in right of way, pavements, traffic volume.

The system was introduced first in 2012 by an external consultant under the supervision of the Ministry's engineers. The Consultant had the responsibility of installing, collecting, and updating the system with all its related data. To keep this system running properly it is necessary to keep the database updated continuously and for that purpose some high-performance equipment is used such as LiDAR (laser cloud point data collection) which is a 360° photogrammetric. In order to clarify the main objective of installing RAMS is to assess the administration to reach a point where the big part of assets management can be achieved and quantified from the preventive perspective (by modeling and forecasting) and maintaining the less of the constituting points of the RAMS on the corrective side, for example, the daily maintenance performed in which the incidents was unexpected) since it shows that on long term, performing preventive actions can lead to the reduction of payment by 3–5 times compared to the corrective works on conventional method.

7 Conclusions

This paper has explored the importance of using recycled asphalt from a wide perspective to ensure excellent quality of infrastructure, especially roads network. Recycled asphalt is a tool that is used during the preventive maintenance process which monitors and evaluates the pavement conditions. This can recognize the level of failures in the current situation and suggest new maintenance forecasting for the following years. This supports the idea of establishing new processes and techniques that can associate new business markets that empower the economy of any country.

References

Koch, S., & Ksaibati, K. (2010). *Performance of Recycled Asphalt Pavement In Gravel Roads.* The University of Wyoming.

Byrne BEng, D. (2005). Recycling of Asphalt Pavements in New Bituminous Mixes. Ph.D. Napier University

Rosyidi, P.S. (2015). Construction Method for Road-Pavement

Asphaltwa.com (2010). Asphalt Pavement History | Washington Asphalt Pavement Association. https://www.asphaltwa.com/welcome-asphalt-pavement-history/. Accessed 1 Nov 2020

Transportation Research Board (2014). Application of Reclaimed Asphalt Pavement and Recycled Asphalt Shingles in Hot-Mix Asphalt. Washington, D.C. 20001: Transportation Research Board, pp. 11–18

Koch, S., Huntington, G., & Ksaibati, K. (2013). *Performance of Reclaimed Asphalt Pavement on Unpaved Roads.* University of Wyoming.

Kandhal, P., Mallick, R. (2017). Chapter 9, 10 - 98042 - Federal Highway Administration. https://www.fhwa.dot.gov/pavement/recycling/98042/10.cfm. Accessed 28 Feb 2022

Lombardo, J. (2018). Should You Use Hot or Cold in Place Recycling for Your Project? For Construction Pros. https://www.forconstructionpros.com/asphalt/article/20984416/should-you-use-hot-or-cold-in-place-recycling-for-your-project. Accessed 2 Apr 2022

Reeder, G., Harrington, D., Ayers, M., & Adaska, W. (2019). *Guide to Full-Depth Reclamation (FDR) with Cement* (pp. 5–11). Iowa State University.

Toolkit.irap.org (2022). Shoulder Sealing | Road Safety Toolkit. https://toolkit.irap.org/default.asp?page=treatment&id=32. Accessed 1 Mar 2022

Balaguera, A., Carvajala, G., Albertí, J. and Palmer, P. (2018). Life cycle assessment of road construction alternative materials: a literature review. 1(3), 37–48

2017. Highway Maintenance Manual. Bureau of Highway Maintenance. Bureau: Bureau of Highway Maintenance, pp.1–5. https://wisconsindot.gov/Documents/doing-bus/local-gov/hwy-mnt/mntc-manual/chapter02/02-10-15.pdf. Accessed 24 Jan 2022

Overview of Concrete Deterioration Due to Sulphate Attack and Comparison of Its Requirements in International Codes

Diala Basim Al-Haddad, Gul Ahmed Jokhio, and Abid Abu Tair

Abstract Any material or geometric property change that has an impact on structural performance is known as structural deterioration. Although this phenomenon is gradual, if it is not evaluated and fixed in a timely manner, it will have a severe impact on the structure and, in some historical occasions, may result in collapse. Deterioration has a variety of causes and mechanisms. Some are regarded as external, such as the environment, which is critical in areas with tough climate conditions, as the Arab Gulf, where temperatures and humidity are high, another factor is the low quality of the materials and poor craftsmanship. Depending on the type of materials used in the structure and the environment in which it is created, several precaution approaches for structural degradation may be adapted.

Sulphate attack is one of the most widespread deteriorating mechanisms; this type of chemical damage can result in cracks, spalling, and disintegration of the structure, diminishing its strength. This paper presents a thorough investigation of this degrading agent, including its causes, effects, international code provisions, as well as recent studies on the subject with a focus on the Arab Gulf Countries.

The outcomes of this study demonstrated that this phenomenon, with its various mechanisms, has a negative impact on structures. Despite the distribution of sulphates in the Gulf, the existence of chloride and the precautionary measures taken successfully limited their impact on structures.

Keywords Sulphate attack · Chemical reaction · Harsh environment · Ettringite · Physical attack

D. B. Al-Haddad (✉)
The British University in Dubai, Dubai, UAE
e-mail: haddad.diala@hotmail.com

G. A. Jokhio · A. A. Tair
Structural Engineering, The British University in Dubai, Dubai, UAE
e-mail: gul.ahmed@buid.ac.ae

A. A. Tair
e-mail: abid.abu-tair@buid.ac.ae

© The Author(s) 2023
K. Al Marri et al. (eds.), *BUiD Doctoral Research Conference 2022,*
Lecture Notes in Civil Engineering 320,
https://doi.org/10.1007/978-3-031-27462-6_19

1 Introduction

One of the most crucial characteristics to consider in structures is their durability. It is critical that structures maintain the attributes for which they were designed for in all the environmental situations where they are employed.

Structures may be subjected to different degradation agents throughout their life cycle, posing a hazard to their long-term sustainability. These agents could be external, because of the structure's environmental influence, or internal, because of the matrix of constituents that make up the structure (Findik & Findik, 2021).

Structural environment plays a vital role in the service life of concrete structures, especially in the areas that have sever climate conditions such as the Arab gulf. These countries are known for their hot-humid or hot-dry weather conditions combined with salt air-contaminated particles that -if not considered- will dramatically affect the durability of the structure (Al-Gahtani & Maslehuddin, 2002).

Mechanism of concrete deterioration might be physical; for example, caused by the discrepancy between the cement paste and aggregate thermal properties. Mechanical, usually related to abrasion, and chemical, caused by the attack of acids, saltwater, sulphates, and other elements (Findik & Findik, 2021).

Sulphate attack is one of the damaging agents for structures. The by-products of this reaction begin to chip away at the paste that maintains the concrete together. New chemicals, such as ettringite, form as sulphate dries. These new crystals fill the gaps in the paste, causing it to break and further damage the concrete.

Sulphate attack causes billions of dollars in damage to concrete such as wastewater collection and treatment facilities all over the world (Pan et al., 2017), thus this phenomenon's mechanism, mitigation, protection of structures during their life time is critical.

2 Sulphate Attack Overview

2.1 Sulphate Attack Definition

Sulphate attack is a degradation process in which sulphate ions strike the constituents of cement paste. Water-soluble sulphate-containing salts such as alkali-earth (calcium, magnesium) and alkali (sodium, potassium) sulphates, which are chemically reactive with concrete components, cause sulphate attack. Sulphate attack on concrete can take several forms, depending on the chemical type of sulphate and the exposure of the concrete to the environment. (See Fig. 1)

Sulphate sources might be internal or external. Internal sources are less common, but they appear with materials used in concrete, such as hydraulic cements, fly ash, aggregate, and admixtures. External sources are more common, typically soils and ground water have high-sulphate content, as well as pollution from the atmosphere or industry (Zhao et al., 2020).

Fig. 1 Structure affected by
sulphate attack. (Suryakanta,
2015)

2.2   Forms and Mechanism of Sulphate Attack

There are two main forms of sulphate attack: chemical or classical form and physical or salt attack. The creation of ettringite and the formation of gypsum are the two principal processes of chemical sulphate attack, according to ACI's Guide to Durable Concrete (201.2R). However, more forms can be identified because of sulphate chemical interactions with cement hydration products. See Fig. 2

Classical (chemical) Form of Sulphate Attack
This mechanism of sulphate attack result in products that have expansive volume which will cause internal tensile stresses on the concrete and cause cracks, spalling, and/or disintegration.

Ettringite Formation
There are two types of Ettringite, the first one is labelled as primary ettringite, formed after few hours of blending water and cement in a termed process of Early Ettringite Formation (EEF), this type doesn't cause any remarkable damage despite of its

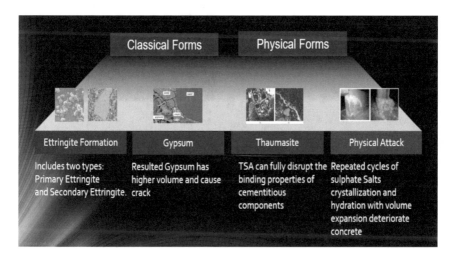

Fig. 2 Mechanism of sulphate attack

big volume, it is consistently and individually distributed at a sub-microscopic level across the cement paste, which functions as a covering over the surfaces of cement grains shortly after mixing, regulate concrete setting. Because of its pore filling function, primary ettringite can raise strength, lower permeability, porosity, and provide dimensional stability.

The other ettringite type is called secondary ettringite and formed in Delayed Ettringite Formation (DEF) process. DEF starts when the primary ettringite generated dissolves at high curing temperatures over 65–70 °C when the temperature of concrete dips below 70 degrees Celsius, the ettringite re-forms to secondary ettringite, causing the hardened concrete to expand and crack.

This type developed in concrete that has been cured at high temperatures (steam curing) and in large pours where the heat of hydration has caused high temperatures within the core area. DEF causes the paste to expand while the aggregate does not, resulting in fractures (or gaps) surrounding the aggregate, with the larger the aggregate, the larger the gap. (See Fig. 3).

Gypsum Formation

Gypsum, often known as calcium sulphate dihydrate, is a soft sulphate mineral having a hardness of 2 on the Mohs scale. (See Fig. 4).

The hydration of Portland cement's silicate phases releases lime. Gypsum is formed when sulphate ions react with calcium hydroxide. This reaction result has a higher solid volume than the original ingredients, which may contribute to concrete degradation in some situations.

Fig. 3 Secondary ettringite causing a gap around aggregate.(Sulphate attack in concrete 2017)

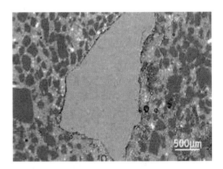

Fig. 4 Gypsum Mineral (Vasavan, 2017)

Fig. 5 TSA affected
concrete.
(Thaumasite-attack. 2022)

Thaumasite form of Sulphate Attack
The Thaumasite form of sulphate attack necessitates the presence of both carbonate and sulphate ions in solution. Thaumasite is formed when calcium silicate hydrate (C-S–H) and calcium hydroxide are degraded by reactions with sulphate attack and carbonate.

For TSA to occur, a low temperature (4–10) °C is required, and severe damage occurs only when the ground is quite damp. TSA is a rather uncommon type of sulphate attack, with little probability of occurrence unless the environment is a combination of cold weather, presence of sulphate ions, carbonates, and mobile groundwater. Concrete becomes a friable material when it hardens. TSA-affected concrete can be easily fractured with fingers, and the coarse aggregate can be removed. (See Fig. 5).

Physical form of sulphate attack
Chemical sulphate attack on concrete structures has long been thought to be the most common cause of concrete deterioration in sulphate-rich environments. However, it was shown that under specific environmental conditions, concrete is primarily affected by physical sulphate damage. (Suleiman, 2014).

Bloom (presence of sodium sulphate salts) at exposed concrete surfaces is a common sign of physical sulphate attack. It's not only an aesthetic issue; it's also a visual sign of chemical and microstructural destruction in the concrete.

Capillary action and diffusion allow sulphate salts to enter the pore spaces of concrete in solution. The wick's motion draws the sulphate solution to the exposed surface, where it evaporates, gradually increasing the sulphate ion concentration until it crystallizes. The sulphate salts go through cycles of crystallization and dissolution, or hydration and dehydration, as the ambient temperature and relative humidity change. Repeated cycles of crystallization and hydration with volumetric expansion can induce concrete deterioration comparable to that caused by freezing-and-thawing cycles. (See figure 6).

2.3 Sulphates are Not Equal

Sulphates can be rather very aggressive according to which cation they are coupled with, moderately aggressive, or low aggressive.

Fig. 6 Physical sulphate
attack on concrete
(Suleiman, 2014)

MgSO4 (magnesium sulphate) is the most aggressive (Rachel Detwiler, 2021) whereas calcium sulphate (CaSO4) is the least destructive since it is the least soluble. Even so, due to the production of ettringite, it will cause expansion and cracking. Sodium sulphate, often known as Na2SO4, is a moderately aggressive substance. Because of the sodium ions, there is enough NaOH in solution to keep the calcium silicate hydrate gel, the cement paste's main strength-producing component, stable. Because it does not contain calcium, unlike CaSO4, it will attack calcium hydroxide. Due to the development of both ettringite and gypsum, it promotes expansion and cracking.

Scholars from early time investigated the factors that affect this detrimental process, for example, Dhole et al. (2019), Qiang Yuan et al. (2021) and Liu et al. (2020) stated that Sulphate attack on concrete is influenced by a number of elements, including the type and quantity of sulphate solution, temperature, pH value, cement composition, admixtures, and erosion form. The most essential elements in sulphate attack are the temperature, concentration, and kind of sulphate solution.

3   Structure Deterioration in the Arab Gulf Due to Sulphate Attack

3.1   Sulphate Sources and Distribution in UAE and Arab Gulf

The existence of sabkha soil in the Arab Gulf is largely responsible for the deterioration, notably of columns and footings.

The geology of the Arabian Gulf coast, including the UAE and several neighboring countries, is dominated by Sabkha, a gypsum-rich deposit formed inland by tidal sea water evaporation. Sabkha's ground water contains gypsum, anhydrite, calcium carbonate, and sodium chloride. Concrete structures constructed in Sabkha are particularly vulnerable to sulphate and chloride attack, increasing the risk of deterioration. (Khan, 2018)

Other than that, Saleh (2014) investigated the amount and distribution of different sulphate types in two areas: Al Khatem and Remah in Abu Dhabi, UAE. The results are shown in the map below. (See Fig. 7). The main source of sulphate here is the volcanic and sedimental rocks.

Fig. 7 Distribution map for sulphate distribution

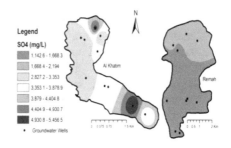

3.2 Current Situation Structure Deterioration due to Sulphate Attack in the Arab Gulf

Even though the weather conditions are harsh and fluctuating and sulphate concentration in the soil and ground water is excessive, no severe concrete deterioration has been reported because of this phenomenon. The use of Type V cement with a low C3A component was most likely to attribute. Additionally, the presence of chloride ions, which protects against sulphate attack limit structure deterioration.

The Yas Island Water project in Sabkha is representative of most projects in the Arabian Peninsula that are exposed to a sulphate and chloride-laden environment. Ground water chemical testing revealed the existence of an aggressive chemical environment due to the inclusion of gypsum layers beneath, accounting for roughly 49% of the oxide element, according to the Geotechnical interpretive report. (Saleh, 2014)

4 Sulphate Attack Requirements in the International Codes

International codes, such as the American code ACI 318, the British standards BS 8500–1, and the European code EC2, provide provisions for the structural design of durability against sulphate attack.

ACI 318–19 classifies concrete in contact with soil or water that contains concentrations of water-soluble sulphate ions as exposure category S. This category is further divided up into four classes, with S0 indicating low levels of water-soluble sulphate. S1, S2, and S3 exposure classes are designated to structural concrete members that have direct contact with soluble sulphates in soil or water. From Exposure Class S1 to S3, the severity of exposure increases. Table 19.3.2.1 shows the maximum allowable w/c ratio, the minimum compressive strength, the cement type and cementitious material, and the permissible use of calcium chloride admixtures (ACI Committee 318 2019).

Similarly, when concrete is exposed to sulphate-containing soil or natural water, the British code exposures include XA1 for a fractionally aggressive environment, XA2 for a moderately aggressive environment, and XA3 for a highly aggressive environment. The maximum allowable w/c ratio, minimum cement or combination content, and indicative strength are also specified in Tables A.4 and A.5. Furthermore, the standard considered the source of sulphate attack from a concrete constituent: aggregate. Sulphate in aggregate (A.7.5) and Alkali-aggregate reaction (A.8.2) (British Standards Institution, 2019).

The Euro code is based on EN 206-1, Table 2. XA1 Chemical environment that is fairly aggressive XA2 chemical environment is moderately aggressive, and XA3 chemical environment is highly aggressive. (EN1992-1–1 2004).

Overall, these codes provide categories for exposure conditions based on sulphate levels, and for each exposure class, critical value limitations are stated. Numerous detailed studies have examined these values and proposed other performance effective factors in resisting sulphate impact such as (Bentivegna et al. 2020) and (Obla et al., 2017).

5 Case study for Deterioration of Structure due to Sulphate Attack - Cheng-Kun Railway Tunnels

Liu et al. (2017) investigated the deterioration of Cheng-Kun Railway tunnels in China. The structure experienced Concrete failure due to the action of sulphates.

5.1 Visual and Field Inspection

The field investigations resulted in the following: (See Fig. 8)

- A 5 mm thick surface layer was removed from the concrete lining. (a, b).
- After the surface layer was peeled, the aggregates of the concrete were seen, and a significant number of white crystals emerged on the concrete surface. (c)
- Exposed and rusted Steel bars. (d)

Fig. 8 Deteriorated concrete in the tunnel

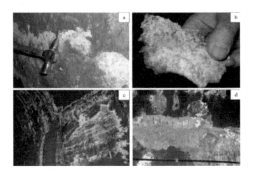

5.2   Test Results

The findings revealed that:

1. Large amounts of Sodium Sulphate were generated causing layer-by-layer detachment.
2. Ettringite and gypsum, were also detected in the neutralized concrete lining.

5.3   Analysis of Findings

Although Sodium Sulphate crystallization caused the concrete to detach, the main product of the detached concrete pieces was Calcium Carbonate. Ettringite and gypsum in the concrete lining indicate that the chemical sulphate attack occurred within the concrete. The concrete lining contained Sodium Sulphate crystals, but it was not detached and was not carbonated with Calcium Carbonate. As a result, it is concluded that chemical sulphate attack occurred in the inner part of the concrete lining, whereas physical sulphate attack occurred on the concrete lining's neutralized surface layer.

5.4   Lessons Learnt and Repair Measures

The study didn't mention any repair plan for the tunnel or the percentage of damage in the tunnel, however, there are some lessons to be learnt and approaches for repair might be considered.

To begin with, this tunnel lining suffered from sulphate ingress from the soil due to the use of normal Portland cement, high w/c ratio, high permeability, and minimal protection, in such cases, where structures are exposed to soil and ground water, sulphate resistance cement together with the lowest possible w/c ratio is preferable.

Depending on the size of the impacted area, two methods can be used for repair: grouting and/or Electrokinetic Nanoparticle Treatment. Grouting is a common practice followed in repairing tunnels.

Sulphate attack is a long process, hence, periodic inspection for any cracks are signs of damage must be rectified. Access points for maintenance and inspection must be part of the design, water proofing membranes would also be beneficial.

6   Discussion and Conclusion

Sulphate attack is harmful to reinforced structures, it has two mechanisms that result in products with high volume: ettringite, gypsum, Thaumasite, and salt crystals.

This enlargement will cause tensile internal stresses in the concrete, causing it to spall, crack, or disintegrate. Furthermore, these cracks may cause steel corrosion by allowing chlorides and other harmful elements to enter. The severity of the damage is determined by the type of sulphate; testing for such elements is required to implement proper mitigation strategies.

Sulphate sources can be internal or external, internal sources coming from concrete composition and external sources from the soil or ground.

International codes, such as ACI, British, and Eurocode, included provisions for durability design to design structures for sulphate ingress resistance. Primarily, they classify sulphate concentrations according to severity and recommend values for strength, concrete cover, water cement ratio, cement type, and SCM.

Sulphate attack can be identified by the following observations: salt crystals on the surface, cracks, and when lightly touching the surface, it quickly scales or flakes away.

Several repair methods can be applied to affected structures depending on the outcome of the attack; for cracks, based on their type, repairing methods can include epoxy injection, routing and sealing, stitching, drilling, and plugging, gravity filling, grouting, and crack overlay and surface treatment. If spalling occurs, new binding materials of the same quality as the old ones be used. Electrokinetic Nanoparticle Treatment is a microstructural technique that combines sulphate extraction with nanoscale pozzolan injection.

Based on this overview, we can conclude that sulphate attack is harmful to concrete structures in all of their forms. Multiple international codes included provisions to guide the design engineer with preventive measures to consider designing against it.

Multiple approaches to limiting SA are available; however, if sulphate attack caused damage, repair methods can be used to restore its effect.

Sulphates are abundant in the Arab Gulf region. Particularly in the Sabkha region, of which the UAE is a part. The awareness of its effect in the UAE, as well as the implementation of design precautions and proper curing, aided in confining its impact, particularly considering the region's harsh environmental conditions.

7   Recommendation

Overall, the following is recommended:

1. Environmental inspection of any sulphate and its concentration must be completed prior to the executive phase of any project and design for durability as per the codes should be considered.
2. A sulphate attack risk mitigation plan must be developed according to code provisions.
3. Regular inspection and correction of any signs of sulphate assault are required.

References

ACI Committee 318. (2019). Building code Requirement for structural concrete ACI318–19.

Al-Gahtani, A. S. & Maslehuddin, M. (2002). Characteristics of the Arabian Gulf environment and its impact on concrete durability-an overview. *The 6th Saudi Engineering Conference*, KFUPM, Dhahran, Saudi Arabia, vol. 3(December), pp. 169–184.

British Standards Institution. (2019). Complementary British Standard to BS EN 206 Part 1: Method of Specifying and Guidance for the Specifier (BS 8500-1:2015+A2:2019 Concrete).

Dhole, R., Thomas, M. D. A., Folliard, K. J., & Drimalas, T. (2019). Chemical and physical sulfate attack on fly ash concrete mixtures. *ACI Materials Journal, 116*(4), 31–42.

EN1992-1-1. (2004). Eurocode 2: Design of concrete structures - Part 1-1 : General rules and rules for buildings. Brussels: EUROPEA~

Findik, F., & Findik, F. (2021). Civil engineering materials. *Heritage Sustainable Development, 3*(2), 154–172. https://doi.org/10.37868/hsd.v3i2.74

Khan, Q. R. (2018). Scholarworks @ UAEU Civil and Environmental Theses Assessment of Groundwater Quality in the Al Khatim and Remah Area of the United Arab Emirates.

Liu, P., Chen, Y., Yu, Z., Chen, L., & Zheng, Y. (2020). Research on sulfate attack mechanism of cement concrete based on chemical thermodynamics. *Advances Materials Science Engineering, 2020*, 1–16. https://doi.org/10.1155/2020/6916039

Liu, Z., Zhang, F., Deng, D., Xie, Y., Long, G., & Tang, X. (2017). Physical sulfate attack on concrete lining–A field case analysis. *Case Studies Construction Materials, 6*(February), 206–212.

Obla, K., Lobo, C., Spring, S. & Kim, H. (2017). An Evaluation of Performance- Based Alternatives to the Durability Provisions of the ACI 318 Building Code, vol. 5(179).

Pan, X., Shi, Z., Shi, C., Ling, T. C., & Li, N. (2017). A review on surface treatment for concrete – Part 2: Performance. *Construction Building Materials, 133*, 81–90.

Yuan, Q., Liu, Z., Zheng, K., & Ma, C. (2021). Inorganic cementing materials. *Civil Engineering Materials* (pp. 17–57). Elsevier. https://doi.org/10.1016/B978-0-12-822865-4.00002-7

Rachel Detwiler. (2021). What is Sulfate Attack? Beton Construction Engineering, pp. 7–8. https://www.slagcement.org/aboutslagcement/is-07.aspx.

Saleh, A. (2014). Development of Sustainable and Low Carbon Concretes for the Gulf Environment: Development of Sustainable and Low Carbon Concretes for the Gulf Environment Abu Sale.

Suleiman, A. (2014). Physical Sulphate Attack on Concrete. University of Western Ontario - Electronic Thesis and Dissertation Repository., vol. Master of(May), pp. 20–58 [online].Available at: http://ir.lib.uwo.ca/etd/2058/.

Sulfate attack in concrete. (2017). Understanding Cement.: https://www.understanding-cement.com/sulfate.html. Accessed 17 May 2022

Suryakanta. (2015). How to protect concrete from sulphate attack? - CivilBlog.Org. Civil blog.org. https://civilblog.org/2015/02/27/how-to-protect-concrete-from-sulphate-attack/#. Accessed 1 May 2022

thaumasite-attack.jpg (503 × 326). (2022). https://taylortuxford.co.uk/wp-content/uploads/2018/03/thaumasite-attack.jpg. Accessed 17 May 2022

Vasavan, M. (2017). Sulfate Attack and Sulfate Content in Concrete: A Literature Review | Manilal Vasavan | Pulse | LinkedIn. LinkedIn, (9), pp. 1–11. https://www.linkedin.com/pulse/sulfate-attack-content-concrete-literature-review-manilal-vasavan.

Zhao, G., Li, J., Shi, M., Fan, H., Cui, J., & Xie, F. (2020). Degradation mechanisms of cast-in-situ concrete subjected to internal-external combined sulfate attack. *Construction Building Materials, 248*, 118683. https://doi.org/10.1016/j.conbuildmat.2020.118683

Printed in the United States
by Baker & Taylor Publisher Services